Springer Theses

Recognizing Outstanding Ph.D. Research

Aims and Scope

The series "Springer Theses" brings together a selection of the very best Ph.D. theses from around the world and across the physical sciences. Nominated and endorsed by two recognized specialists, each published volume has been selected for its scientific excellence and the high impact of its contents for the pertinent field of research. For greater accessibility to non-specialists, the published versions include an extended introduction, as well as a foreword by the student's supervisor explaining the special relevance of the work for the field. As a whole, the series will provide a valuable resource both for newcomers to the research fields described, and for other scientists seeking detailed background information on special questions. Finally, it provides an accredited documentation of the valuable contributions made by today's younger generation of scientists.

Theses are accepted into the series by invited nomination only and must fulfill all of the following criteria

- They must be written in good English.
- The topic should fall within the confines of Chemistry, Physics, Earth Sciences, Engineering and related interdisciplinary fields such as Materials, Nanoscience, Chemical Engineering, Complex Systems and Biophysics.
- The work reported in the thesis must represent a significant scientific advance.
- If the thesis includes previously published material, permission to reproduce this must be gained from the respective copyright holder.
- They must have been examined and passed during the 12 months prior to nomination.
- Each thesis should include a foreword by the supervisor outlining the significance of its content.
- The theses should have a clearly defined structure including an introduction accessible to scientists not expert in that particular field.

More information about this series at http://www.springer.com/series/8790

Sownak Bose

Beyond ΛCDM

Exploring Alternatives to the Standard Cosmological Paradigm

Doctoral Thesis accepted by
the University of Durham, Durham, UK

 Springer

Author
Dr. Sownak Bose
Harvard–Smithsonian Center
  for Astrophysics
Cambridge, MA, USA

Supervisors
Prof. Carlos S. Frenk
University of Durham
Durham, UK

Dr. Baojiu Li
University of Durham
Durham, UK

Prof. Adrian Jenkins
University of Durham
Durham, UK

ISSN 2190-5053   ISSN 2190-5061   (electronic)
Springer Theses
ISBN 978-3-030-07245-2   ISBN 978-3-319-96761-5   (eBook)
https://doi.org/10.1007/978-3-319-96761-5

This Springer imprint is published by the registered company Springer Nature Switzerland AG
The registered company address is: Gewerbestrasse 11, 6330 Cham, Switzerland

There are only two problems with ΛCDM:
Λ, and CDM.

—Tom Shanks

Supervisors' Foreword

Sownak Bose's Ph.D. thesis is exceptional: it is effectively two theses in one, on two different topics. One is the formation of cosmic structure in models in which the dark matter is made of "warm" elementary particles (WDM) rather than the "cold" particles (CDM) assumed in the standard cosmological model (known as ΛCDM). The other is modified gravity, an attempt to explain the accelerated expansion of the Universe by extending Einstein's theory of General Relativity.

Using the highest resolution simulations to date of cosmological regions in WDM and CDM universes (the "Copernicus Complexio" or COCO simulations), Sownak carried out the most comprehensive and rigorous comparative study so far of the formation, evolution, abundance, and internal structure of dark matter halos and subhalos in WDM and CDM. The results were presented in two papers he led, both published in 2016, which are likely to become the standard reference in the subject. He then examined whether the epoch of reionization—when the Universe became transparent to light as the primordial hydrogen was ionized by the first stars—might be a discriminant between the two models. For this he applied Galform, a comprehensive semi-analytic model of galaxy formation, to COCO. Surprisingly, he found that there is little difference in the epoch when the universe is reionized in the two cases. This is because, as he showed, the first objects that form in WDM are already relatively massive and very efficient at producing UV photons.

Sownak made major contributions to other studies of the nature and evolution of cosmic structure in WDM universes that are reported in his thesis. One of the chapters deals with an important difference between WDM and CDM dark matter halos. The "initial conditions" for the formation of structure in WDM models contains a preferred scale that reflects the "warmth" (or initial thermal velocities) of the WDM particles. This scale is reflected both in the growth path (or "mass accretion history") of dark matter haloes and in the relation between the final mass and concentration of the halo. Working with A. Ludlow and others, Sownak developed an elegant formalism to include the effect of a preferred scale on the structure of the halo and generalized it to include also the case of CDM. This work too has been influential and the recent paper in which it is described is already highly cited.

In a particularly attractive version of WDM, the dark matter consists of particles called sterile neutrinos with a mass of a few kilo-electronvolts. Working with M. Lovell and others, Sownak analyzed state-of-the-art cosmological simulations which follow not only the evolution of these dark matter particles but also the evolution of the ordinary (or baryonic) matter that makes up the visible Universe. A defining characteristic of WDM (related to the scale mentioned earlier) is that it gives rise to far fewer small dark matter halos than CDM where no such scale is present. In the real universe, galaxies in small halos are so faint that they can only be seen as satellites orbiting around our Milky Way galaxy. This chapter of the thesis shows that a subset of sterile neutrino models can be discarded because they fail to produce enough halos to account for the observed number of satellites but a subset, with a different parameter, gives an excellent match to the observed luminosity function of satellites (as does CDM). In future, it may be possible to distinguish CDM from WDM by analyzing images of a particular type of gravitational lensing phenomenon knows as "Einstein rings." Their images can be distorted by intervening small-mass halos whose numbers differ greatly in WDM and CDM. In a paper led by R. Li, Sownak contributed to a detailed calculation of how this test would work.

Sownak's research on CDM and WDM alone would have made an outstanding thesis. However, Sownak also worked on modified gravity theories, a topic in which he co-authored five papers as a student. First, he focused on a class of particularly popular modified gravity theories, called $f(R)$. Most studies make use of a quasi-static approximation even in the highly nonlinear regime appropriate to galaxies and clusters. Sownak managed to validate this approximation which constitutes an important fundamental check without which the results of full cosmological simulations of these models cannot be trusted. He also worked mainly on the development of methodology in which he achieved an important breakthrough: the invention of an algorithm to speed up simulations for the major classes of models, including $f(R)$ gravity. This was built upon his observation that the previous, widely used, method made the relevant equations highly nonlinear and, by making a variable redefinition, he brought the equations into a much less nonlinear form that can be solved analytically; this led to over an order-of-magnitude improvement in the efficiency of the simulation.

In total, Sownak's thesis produced 17 refereed papers, in 5 of which he is first author. It is one of the broadest, most comprehensive, and technically accomplished theses out of the 35 that I have supervised during my career. And it has also had significant impact in the subject as evidenced, for example, by the over 400 citations that the papers in the thesis, all published in the past 3 years, have so far accrued.

Durham, UK   Prof. Carlos S. Frenk
May 2018

Preface

Understanding the nature of the dark matter and dark energy is perhaps the most outstanding problem in modern cosmology. While their important role in the evolution of the Universe has been well established—that dark matter acts as the building blocks of galaxies and that dark energy accelerates the expansion of the Universe—details on their true nature remain elusive.

Astronomers often quip that placing the word "dark" in front of a cosmological term simultaneously describes both its unobserved nature and our lack of understanding of it. As the standard model of cosmology has been established, we have come to appreciate that Nature is somewhat cruel: while the "ordinary", known species of matter make up only around $\sim 5\%$ of the Universe, the remaining bulk of the cosmos is composed of the mysterious, invisible dark matter, and dark energy.

Dark matter, as its name suggests, is hypothesized to be matter (likely a fundamental particle) that interacts extremely weakly (if at all) with photons, allowing structures to form through gravitational clumping into deep potential wells. Dark energy is perhaps even stranger: believed to be an all-pervading "force" that counteracts gravity on cosmological scales, driving the accelerated expansion of the Universe. Historically, astronomy has been the study of the cosmos with the use of light received in telescopes; in a funny way, therefore, the study of dark matter and dark energy is perhaps the very antithesis of traditional astronomy.

Over the past three decades or so, a new program of investigation has become very popular: the use of numerical simulations to test models of structure formation made possible by the increasing power and availability of computing algorithms and architectures. Numerical simulations have played a key role in furthering our understanding of the unobserved Universe, and to better comprehend the physical processes associated with the formation of stars and galaxies.

The content of this thesis is split into two parts where we consider, in turn, alternatives to the two main constituents of the standard model of cosmology: *cold dark matter* (CDM) and the *cosmological constant*, Λ. In particular, we perform detailed simulations of "artificial universes" with these alternative models. In the first half, we consider "sterile neutrino" dark matter, motivated by claims that these particles may have been detected recently through their decay in clusters of galaxies. With sterile neutrinos as the dark matter, structure formation is delayed and the abundance of dwarf galaxies is suppressed relative to a CDM universe. Using sophisticated models of galaxy formation, we find that future observations of the early Universe and of faint dwarf galaxies in the Local Group can place strong constraints on the sterile neutrino scenario.

In the second half, we propose and test novel numerical algorithms for simulating Universes with a "modified" theory of gravity: where we retain CDM, but replace Λ with small modifications to the Einstein field equations of General Relativity. The interest in modified gravity theories is not limited to just trying to explain the observed acceleration of the Universe, but also to quantify and test the enhancement of gravity (via a "fifth force") on cosmological scales. Testing the nature of gravity on cosmological scales is one of the primary science objectives of a number of present and upcoming surveys; for this reason, it is particularly timely to consider the possible deviations from General Relativity that are predicted by alternative theories of gravity. The focus of this thesis is on the $f(R)$ model, which is one of the most widely studied theories of modified gravity. The numerical techniques elucidated in this thesis improve the efficiency of these simulations by more than a factor of 20 compared to previous methods. These new methods will be something of a breakthrough for modified gravity simulations, opening up a path toward "precision cosmology" tests of gravity that would have been difficult to achieve previously. Furthermore, it will now become feasible, for the first time, to couple these theories to sophisticated treatments of galaxy formation physics.

In Chap. 2, we introduce the *Copernicus Complexio* simulations (COCO; Hellwing et al. 2016; Bose et al. 2016) and compare the properties of sterile neutrino dark matter haloes to CDM haloes. We establish and compare the mass function of haloes, as well as their structural properties such as density profiles, the mass-concentration relation, shapes, and spins. In Chap. 3, we focus on the properties of substructures in coco, such as their abundance, radial distribution within host haloes, and the effects of tidal stripping on these substructures. We then run the Durham semi-analytic galaxy formation model, galform, on our simulations to investigate differences in the galaxy population between the sterile neutrino and CDM simulations. In Chap. 4, we use galform to investigate if the epoch of reionization and the present-day abundance of Milky Way satellites can be used to constrain a range of 7 keV sterile neutrino models.

In the second part of the thesis, we shift our focus to modified gravity theories, using the well-known $f(R)$ model as the working example. In Chap. 5, we test the validity of the widely used *quasi-static approximation*, which assumes that any time variation of the $f(R)$ gravity scalar field is negligible compared to its spatial variation. In Chap. 6, we present and demonstrate a new and efficient method for simulating certain classes of modified gravity theories, once again using $f(R)$ gravity as the representative example. The results of Chaps. 5 and 6 will hopefully elevate future numerical simulations to a level that will allow for precision cosmological tests of modified gravity.

Finally, in Chap. 7, we summarize the results of this thesis and discuss future research avenues for constraining models beyond ΛCDM.

Cambridge, MA, USA   Dr. Sownak Bose
May 2018

References

Bose S, Hellwing WA, Frenk CS, Jenkins A, Lovell MR, Helly JC, Li B (2016) MNRAS 455:318
Hellwing WA, Frenk CS, Cautun M, Bose S, Helly J, Jenkins A, Sawala T, Cytowski M (2016) MNRAS 457:3492

Declaration

The work described in this thesis was undertaken between 2013 and 2017, while the author was a research student under the supervision of Prof. Carlos S. Frenk, Dr. Baojiu Li, and Prof. Adrian Jenkins at the Institute for Computational Cosmology in the Department of Physics at the University of Durham. No part of this thesis has been submitted for any other degree at the University of Durham or any other university.

The contents of this work have appeared in the following papers:

- **Bose, S.**, Hellwing, W. A., Frenk, C. S., et al., "The Copernicus Complexio: statistical properties of warm dark matter haloes", 2016, MNRAS, 455, 318 (Chap. 2).
- **Bose, S.**, Hellwing, W. A., Frenk, C. S., et al., "Substructure and galaxy formation in the Copernicus Complexio warm dark matter simulations", 2017, MNRAS, 464, 4520 (Chap. 3).
- **Bose, S.**, Frenk, C. S., Jun, H., Lacey, C. G., Lovell, M. R., "Reionization in sterile neutrino cosmologies", 2016, MNRAS, 463, 3848 (Chap. 4).
- **Bose, S.**, Hellwing, W. A., Li, B., "Testing the quasi-static approximation in $f(R)$ gravity simulations", 2015, JCAP, 2, 034 (Chap. 5).
- **Bose, S.**, Li, B., Barreira, A., et al., "Speeding up N-body simulations of modified gravity: chameleon screening models", 2017, JCAP, 2, 050 (Chap. 6).

Elements of Sects. 2.6, 3.5, and 7.2.2 have appeared in

1. Ludlow, A. D., **Bose, S.**, Angulo, R. E., et al., "The mass-concentration-redshift relation of cold and warm dark matter haloes", 2016, MNRAS, 460, 1214
2. Li, R., Frenk, C. S., Cole, S., Gao, L., **Bose, S.**, Hellwing, W. A., "Constraints on the identity of the dark matter from strong gravitational lenses", 2016, MNRAS, 460, 363
3. Lovell, M. R., **Bose, S.**, Boyarsky, A., et al., "Properties of Local Group galaxies in hydrodynamical simulations of sterile neutrino dark matter cosmologies", 2017, MNRAS, 468, 4285

The author has closely collaborated on all three papers. All figures presented in this thesis were prepared by the author, with the exception of Fig. 2.15, which was taken from Ref. [1], Fig. 3.13, which was taken from Ref. [2], and Fig. 7.1, which was taken from Ref. [3].

Acknowledgements

To my mentors. Carlos, Baojiu, and Adrian: thank you for your enduring support, patience, and insight over the past three and a half years—particularly at times when the gaps in my knowledge might have made you question the wisdom in hiring me as your student. In addition to your astronomical expertise, I could not have hoped for a better combination than Carlos's insuppressible optimism, Baojiu's around-the-clock email responses, and Adrian's numerical wizardry. A gentleman by the name of Wojtek deserves a special mention for not only being a great collaborator but also a continuous source of support and, most importantly, for being a good friend. I am also grateful to the many people I have worked with over the past few years, without whom much of this thesis would not be possible. My thanks to my former MPhys supervisor, Yang-Hui He, for being my first mentor in the world of academic research and exposing me to the joys of theoretical physics. Finally, I would like to express my gratitude to Lydia, Alan, and John for never (openly) berating me for my incessant computing queries, and to Sabine, Lindsay, and Dorothy, for an infinite number of reasons.

To my benefactors. The Science and Technology Facilities Council (STFC): for funding me throughout the duration of my Ph.D., ensuring that I haven't needed to resort to any illicit activities to keep myself fed and watered.

To my friends in the department. Andrew, Helen, Campbell, Paddy, Tim, Saavi, Hou Jun, Trayford, Hannah, Matthieu, Alex B., Peter, Flora, Collinson, Tamsyn, Charles, Michelle, Jascha, Violeta, Agnese, Aaron, Alejandro, Marius, Nuala, Cooper, Yang-Shyang, Claudio, Alasdair, Bitten, Julie, Oliver, Anna, Griffin, Ruari, Alex S., and Stefan: your individual and collective presence has always made work such an enjoyable place to be. Thank you to Mathilde for being my constant companion in exhausting on-demand televisual entertainment. And of course, a necessary shout out to my various office mates, past and present: the old R303 C&C Wednesday crew (Cowley, George, Michael, and Jan), and my last set of neighbors in Stuart, Jaime, and Piotr. I am grateful that you never ratted me out when I've just sat around watching football in the office.

xvi Acknowledgements

To my friends outside the department. Justin, Mika, Claire, and Chris: for being great housemates away from the rock 'n' roll lifestyle of astrophysics, and for sharing innumerable great memories forged in the wastelands of Flass Street. And Jon too, I suppose, for being generous enough to not drive me *completely* insane before the end. Everyone in The Three Tuns ca. 2013/14, Ellinor, Louise, Ginny, Anabel, Bill, Rick, Michelle, J-Nad, Kayla, Simone, and anyone I may have missed. What a dreary experience Durham might have been without you all! I will miss those faces from Freeman's Quay that became so familiar over the past 3 years, and which will forever be associated with great physical pain and mental exhaustion. I am also indebted to Falafel Alhana for providing an ever-reliable source of nutrition for the past 3 years. Well, reliable until they stopped doing parmos.

To my friends from a former life. Dubai, London, Oxford: far too many to name individually, but a special mention to my friends from my undergraduate days for being a part of some of the happiest days of my life.

To my family. My parents: without whom I would not be born, and without whose support and guidance I would be a lesser human being today. My brother: "obviously, yeah" for being the world's first privately educated roadman.

To the great love of my life. The dictator of my mood; my constant through thick and thin. For as long as I can remember, we have shared good times and bad, interspersing sadness with optimism. I wouldn't exchange our collective experiences for a thing in the world, and I look forward to the many years together we have ahead of us. Thank you, Manchester United.

Contents

List of Figures

List of Tables

Chapter 1
Introduction

If one describes cosmology as the study of the Universe – its origin, evolution and eventual fate – the conception of this subject can probably be traced back to the earliest annals of human history. Oft-repeated fundamental questions of cosmology such as "Why are we here?" or "How does the Universe work?" put an almost metaphysical spin on what has, over time, become a precision science. It is perhaps due to its dual nature – treading a fine line between science and philosophy – that cosmology has become a subject that has fascinated mankind for millennia. Starting with early records in the Vedic *Rigveda* (ca. 12th century BCE) that describe the Universe as a 'cosmic egg', cycling eternally between periods of expansion and collapse, shifting to the Ptolemaic view (2nd century CE) of an Earth-centred universe, early cosmological models have ranged from themes of the theological to the anthropocentric. The evolution of cosmology from a speculative enterprise to a scientific discipline was made possible through the increasing availability of astronomical data. Following the first recorded 'extragalactic' observations of the Andromeda galaxy made by Persian astronomers (al-Sufi, c.a. 964 CE), the subject of cosmology has undergone a series of metamorphoses, spearheaded by the likes of Copernicus, Galileo, Kepler and Newton (16–17th century CE). Over the course of the 20th century, a standard paradigm has emerged that has not only opened up a wealth of new lines of enquiry into the fundamental questions of cosmology, but has also required a dramatic reassessment of the constituents originally believed to make up our Universe.

1.1 ΛCDM: The Emergence of a Standard Model

While the theoretical groundwork was laid by Einstein in his theory of General Relativity (GR), the kickstarter for the current standard model of cosmology must surely originate from the first extragalactic distance measurements made in the early 20th century. The observed redshifting of spectral lines in 'extragalactic nebulae' (Slipher 1915; Hubble 1929) provided evidence in favour of an expanding Universe, for which the initial condition is a singularity at $t = 0$. In the *Hot Big Bang* scenario,

© Springer International Publishing AG, part of Springer Nature 2018
S. Bose, *Beyond ΛCDM* , Springer Theses,
https://doi.org/10.1007/978-3-319-96761-5_1

the Universe was smaller and denser in the past, as well as being much hotter then than it is now; the composition of the early Universe is therefore believed to be a tightly-coupled sea of photons, electrons and quarks, with this primordial plasma cooling down as the Universe continued to expand. The Hot Big Bang picture is highly predictive: most chiefly, in the form of the background radiation.

Approximately 380,000 years after the Big Bang, expansion cooled the Universe to a temperature at which electrons and protons were able to combine to form hydrogen at an epoch called *recombination*. After this time, the Universe became transparent to photons, allowing them to stream out of the primordial plasma whilst retaining a memory of the initial composition of the Universe. This radiation, redshifted to microwave frequencies by the expansion of the Universe, was detected by Penzias and Wilson in 1964, and was hailed as a monumental discovery in establishing the Hot Big Bang model as the standard paradigm. Due to the all-pervading, isotropic nature of this relic radiation, it has become known as the *cosmic microwave background* (CMB). CMB experiments, particularly in the last two decades, have become extremely valuable for extracting cosmological information about the primordial state of the Universe. The measurement of tiny temperature fluctuations (of the order of $\Delta T / T \sim 10^{-5}$) in the CMB sky by the COBE satellite (Smoot et al. 1992) enabled the measurement of cosmological parameters with unprecedented precision. As subsequent CMB experiments such as WMAP (e.g. Spergel et al. 2003) and Planck (Planck Collaboration et al. 2014) have improved in terms of both sensitivity and angular resolution, the measurement of cosmological parameters has also become ever more precise.

1.1.1 Dark Matter

At around the same time as when observational evidence for the expanding Universe was beginning to accumulate, a mysterious discovery was made by the Swiss astronomer Fritz Zwicky. In his measurements, the virial mass of the Coma cluster, as determined by the observed velocity dispersion of galaxies in the cluster, was estimated to be $\sim 400 \times$ larger than the mass inferred purely from the luminous stellar component (Zwicky 1933). Zwicky's audacious suggestion – that the majority of the mass of the cluster must exist in a 'dark', non-luminous component – received further support following the measurement of flat rotation curves in the outskirts of late-type galaxies (e.g. Babcock 1939; Rubin and Ford 1970), pointing to the existence of a roughly linearly increasing mass profile for such galaxies, far beyond the faint tail of their surface brightness profile.

The requirement for dark matter is further realised through a phenomenon known as *gravitational lensing*. A prediction from GR, lensing asserts that the trajectory of light rays is perturbed in the presence of matter by an amount that is proportional to the amount of intervening matter between the source and an observer. This can be observed as a distortion of the images of background galaxies into characteristic lensing arcs around clusters in the foreground. Analysis of these distortion maps yields a measurement of the mass distribution of the cluster (e.g. Taylor et al. 1998), which again has been found to exceed the mass calculated from the stellar profile alone (e.g. Tyson et al. 1990), hinting at the presence of an additional matter component (though a fraction of the 'missing' mass exists in the form of hot, X-ray emitting gas e.g. Forman et al. 1979, Fabian et al. 1986).

The temperature anisotropies of the CMB constrain the contribution of matter to the total energy density of the Universe to $\sim$30%, of which only $\sim$5% is in the form of the known baryonic matter. The implication that the remaining 25% is in the form of a non-baryonic component is tantalising; investigations into determining its nature has been the focus of a significant amount of theoretical and observational work over the past 30 years. Assuming that the dark matter is a fundamental particle, a natural first candidate to consider was the neutrino. Owing to their very small rest mass (of the order of $\sim$100 eV), neutrinos travel at relativistic velocities, and are thus able to free stream out of small-scale perturbations, erasing fluctuations smaller than the size of superclusters with mass $\sim$$10^{15} M_\odot$. Superclusters are therefore the first structures to form in this *hot dark matter* (HDM) universe; smaller galaxies form as a result of the fragmentation of these larger systems. This 'top-down' nature of structure formation is in direct contradiction to what is observed in the real Universe, putting the HDM interpretation in significant tension with the data. A further hammer blow to HDM was dealt when numerical simulations showed that the large-scale clustering of matter in these neutrino-dominated universes was very different to the clustering observed in the CfA redshift survey (White et al. 1983). For these reasons, HDM soon fell out of favour.

The other limiting case one could consider is where the dark matter is much 'heavier', with a rest mass of the order of a few GeV. In the case of this *cold dark matter* (CDM), the dark matter is assumed to be composed of weakly-interacting massive particles (WIMPs; Peebles 1982), with negligible[1] primordial thermal velocities. As there is now power even on very small scales in CDM, the build-up of structure proceeds *hierarchically* (bottom-up), with larger objects being formed via the merger of smaller clumps. Numerical experiments subsequently demonstrated a remarkably good match between the large-scale clustering of galaxies in a CDM universe with that seen in the CfA redshift survey (Davis et al. 1985). As the size and sophistication of both simulations and data have improved over the past two decades, the CDM model has been rigorously tested over a wide range of scales, and it has, for the most part, passed these tests with flying colours. Cold dark matter has therefore established itself within today's standard model of cosmology.

1.1.2 Dark Energy

Through the 1980s and early 1990s, several theoretical arguments pointed in the direction of another uncomfortable realisation. Inflation strongly suggested a flat

---

[1] 'Negligible' in this context is in comparison to the velocities imparted on the dark matter by gravitational instability.

geometry for the Universe, requiring a total energy density far in excess of that contributed by matter (dark and baryonic) alone. At the close of the 20th century, observations of Type Ia supernovae in distant galaxies (Riess et al. 1998; Perlmutter et al. 1999) provided evidence for a Universe that not only expands, but does so at an *accelerating* rate. It was posited that this accelerated expansion could be generated by the extra energy component required to close the Universe to a flat geometry i.e., contributing the remaining $\sim70\%$ of the energy density of the Universe after accounting for all matter. In the concordance model, the dark energy is sourced by the vacuum, appearing on cosmological scales in the form of a cosmological constant, Λ.

Taken together, the two mysterious components – namely, dark matter and dark energy – that are believed to dominate the total energy density of our Universe, are two major pillars of the current standard model of cosmology, ΛCDM. The most startling fact about this is that despite how little is really known about the properties of the dark matter or the nature of the vacuum energy, predictions of the ΛCDM model have been very successful at fitting cosmological data. Together with the proposition of a Hot Big Bang (and its own predictions such as the CMB and the synthesis of light elements), cosmologists have developed a fairly coherent and complete description of the makeup and evolution of the Universe.

1.2   So, Why Consider Alternatives?

As we have mentioned in the previous section, the concordance ΛCDM model has met with great success in agreeing with observational data spanning a wide range of scales. A valid question to then ask is if there is a need to consider alternatives, to either Λ or CDM, at all – which, indeed, is the subject of this thesis.

1.2.1   Beyond Cold Dark Matter

Arguments against CDM are often presented in the context of the so-called 'small-scale crises' in the model. Specifically, this is in reference to the apparent inconsistencies between the predictions of the CDM model from dark matter-only numerical simulations and what is observed in the properties and abundances of satellites in and around the Local Group. The most famous amongst these, known as the *Missing Satellites* problem, states that the number of dwarf galaxy scale dark matter haloes produced in CDM simulations far outstrip the number of dwarf galaxies actually observed in the Local Group (Kauffmann et al. 1993; Klypin et al. 1999; Moore et al. 1999).

A second discrepancy between theory and observation, this time with regards to the internal structure of satellite galaxy haloes, was described by Boylan-Kolchin et al. (2011). These authors noted that CDM simulations of galactic haloes produced multiple satellites with subhaloes of high internal density (as measured by their peak circular velocity, V_{max}) that had no counterparts in the data. This claim, dubbed the *Too Big to Fail* problem, is similar in spirit to the long-standing *cusp-core* problem (e.g. de Blok et al. 2001), in which the inner slope of the dark matter density profile of satellite galaxies, as inferred from their internal kinematics, is measured to be flatter than the steep inner slope predicted by collisionless, dark matter-only simulations. Alternative models like *warm dark matter* (WDM) have been proposed as potential solutions to these problems, as WDM characteristically reduces the abundance and internal density of dark matter haloes at precisely the scale of interest (i.e., dwarf galaxies), whilst retaining exactly the same behaviour as in CDM on larger scales, where it has been shown to be in good agreement with the observations.

The problem with this way of motivating alternative models to CDM is that the claims have been made using dark matter-only simulations i.e., neglecting the impact of baryon physics on the structure of dark matter haloes, and the likelihood of them hosting galaxies. Hydrodynamical simulations that self-consistently treat processes like reionisation, supernovae and AGN feedback are required to make the most realistic comparisons between observation and theory. Recent works by e.g. Governato et al. (2012), Brooks and Zolotov (2014), Oñorbe et al. (2015), Sawala et al. (2016) have offered solutions to the aforementioned small-scale problems within the context of CDM, without needing to invoke new or exotic dark matter physics.

In fact, the biggest 'problem' with the CDM model is quite simply that, so far, the particle has evaded detection. One of the prime CDM candidates, the neutralino, is the lightest stable particle predicted by supersymmetry. It was hoped that such a particle would be detected in the Large Hadron Collider, yet no evidence for supersymmetry has been found so far. While it has been claimed that CDM particles may have already been observed via their annihilation as an extended gamma ray emission in the Galactic Centre (e.g. Hooper and Goodenough 2011), the dark matter interpretation of this signal is questionable; for example, the contribution of millisecond pulsars or other astrophysical sources to this signal is unclear (e.g. Abazajian 2011; Cholis et al. 2015). While the CDM particle remains undetected, therefore, it is worthwhile to explore viable, well-motivated alternatives.

The alternative we investigate in the first part of this thesis is the case where the dark matter is assumed to be in the form of *sterile neutrinos* (Dodelson and Widrow 1994). The existence of these particles was originally motivated by particle physics: if the Standard Model is extended by adding three right-handed sterile neutrinos, the mixing between sterile and active neutrinos can be used to explain neutrino flavour oscillations. Furthermore, when the masses of these sterile neutrinos are chosen to be below the electroweak scale, it is possible to also account for the asymmetry between matter and antimatter (e.g. Asaka and Shaposhnikov 2005). From a cosmological perspective, the most interesting facet of this model is that the lightest of the triplet of sterile neutrinos with $\mathcal{O}$(keV) rest mass can behave as a WDM particle.

The sterile neutrino interpretation of dark matter has received something of an impetus in the past few years, after the claimed detection of an unidentified emission at 3.5 keV in the stacked X-ray spectrum of galaxy clusters and the Andromeda galaxy (Bulbul et al. 2014; Boyarsky et al. 2014). The line has since also been detected in the Galactic Centre (Boyarsky et al. 2015), the Perseus cluster (Urban et al. 2015) and, most recently, in the Cosmic X-ray Background (Cappelluti et al. 2017), adding weight to the original claims. Just like the case of the gamma ray excess for CDM, it is unclear as to whether or not the emission has an astrophysical origin (see, e.g. Malyshev et al. 2014; Jeltema and Profumo 2015; Anderson et al. 2015; Riemer-Sørensen 2016, for alternative explanations). Recently, Jeltema and Profumo (2016) failed to detect any excess at 3.5 keV in a deep *XMM-Newton* observation of the dwarf spheroidal galaxy Draco, attributing the original line detection to an excitation of K VIII.

If, however, the 3.5 keV line is not of astrophysical origin or an instrumental aberration, it could possibly correspond to the decay of a sterile neutrino with a rest mass of 7 keV.

1.2.2   Beyond Λ

The situation with the cosmological constant, Λ, is a little more complicated. The most severe challenge in associating the vacuum energy as the source of the dark energy is known as the *fine-tuning* problem. This is in reference to the observed value of Λ, inferred from cosmology, which is smaller than the zero-point energy density of the vacuum by at least 60 orders of magnitude. A cancellation of so many powers of ten, which is needed to reconcile the cosmological value of Λ with its quantum mechanical value, requires a fine-tuning mechanism that cannot at the moment be addressed in the Standard Model of particle physics.

A second issue concerns not only the value of Λ, but why it has only just begun to dominate the energy density of the Universe at the present epoch, in what is known as the *coincidence* problem. Denoting the cosmological scale factor by a, where $a = 1$ corresponds to the present day, when $a \ll 1$ (distant past), the matter density dominates Λ, whereas when $a \gg 1$ (distant future), Λ dominates over matter. The fact that we exist at a 'special' time when the relative energy densities of matter and the cosmological constant are roughly comparable (to within an order of magnitude) points to an unlikely coincidence. To reconcile these challenges, anthropic reasoning has often been invoked, suggesting that the existence of intelligent life or the progress of galaxy formation *necessitates* roughly equal contributions of matter and the cosmological constant to the total energy density (e.g. Barrow and Tipler 1986; Weinberg 1987).

Without appealing to some new kind of symmetry, the fine-tuning problem is difficult to address even in most commonly-cited theories of gravity that extend beyond GR plus a cosmological constant. However, these *modified gravity* models, which alter the general relativistic force law beyond some model-dependent scale, are worth investigating primarily because the most rigorous tests of GR have been made within the extent of the Solar System, but not much beyond that. In fact, one of the primary objectives of many upcoming surveys like DESI (Levi et al. 2013) and EUCLID (Laureijs et al. 2011) is to extend these tests of the nature of gravity to larger scales i.e., to look for possible deviations from Λ either in the form of an evolving equation of state of dark energy, or modifications to the gravitational force law itself. In anticipation of the vast amount of data that will soon become available, it is fruitful to quantify any such departures from GR in models that are representative of whole classes of modified gravity theories.

1.3 Tools for Modern Cosmology

Over the past three decades, numerical simulations have played an increasingly prominent role in advancing our knowledge of structure formation in the Universe. The philosophy underlying the numerical method is to simulate a representative patch of the Universe using N discrete point particles to sample the phase-space of the matter field. Given a set of initial conditions – i.e., a set of starting positions, velocities and masses for the particles – the evolution of matter in the simulation volume is tracked by integrating the trajectories of these particles according to Newtonian equations of motion, embedded within a cosmological background.

The dawn of numerical cosmology can be traced back to the pioneering works of von Hoerner (1960), Aarseth (1963) and Peebles (1970), which focussed on the formation and evolution of galaxy clusters using $N = 25$–300 particles. Since then, the availability of larger and more powerful computers, as well as the development of more efficient algorithms for computing gravity have allowed for a tremendous increase in both the size and resolution of N-body simulations. A notable recent example is that of Potter et al. (2016), where the authors reported the completion of a simulation with 2 trillion resolution elements, making it the largest cosmological simulation currently available.

1.3.1 Initial Conditions

The first step in running a cosmological simulation involves setting up its initial conditions (ICs). Creating an accurate set of ICs is fundamental to the final outcome, as even small errors present at early times can be amplified by growing modes over the course of the simulation, and may significantly influence the final result. The simplest particle load that can be generated for the ICs is one where the particles are distributed uniformly (e.g. in a grid configuration). The particles are then perturbed from their initial positions and assigned velocities using e.g. the Zel'dovich approximation (Zel'dovich 1970), which at intermediate and large scales works well enough while density fluctuations are still in the linear regime (high redshift, $z \gtrsim 50$).

Often, one would like to select objects of interest from a large volume and study their structure at much higher resolution. For this purpose, one may employ the 'zoom' technique (Katz and White 1993). Briefly, this technique identifies particles in a sufficiently large volume around the object of interest in the parent simulation, and traces them back to their corresponding location in the unperturbed Lagrangian region. In the re-simulation, this Lagrangian patch is populated with many more particles, each with a smaller particle mass than in the parent simulation, thereby achieving higher resolution in the region of interest. The remaining mass of the cosmological box is sampled coarsely with 'heavy' particles, so as to recover the same large-scale tidal field surrounding the new high resolution region as in the parent simulation. With the added resolution, the new particle load can be perturbed with shorter wavelength Fourier modes than in the parent simulation, adding more power on the small-scales that are now resolved. The zoom technique therefore enhances resolution where it is desired, and compromises by sacrificing resolution elsewhere so as not to dramatically increase the computational cost of the re-simulation.

1.3.2   N-Body Codes

Once the ICs are in place, N-body codes are required to solve the equations of motion, compute the accelerations on the particles, and then integrate their orbits over multiple timesteps. In this thesis, we make use of two state-of-the-art simulation codes: GADGET-3 (Springel et al. 2008, based on the publicly available GADGET-2 code, Springel 2005) and ECOSMOG (Li et al. 2012, based on the publicly available RAMSES code, Teyssier 2002). The two codes follow different philosophies to solving gravity: while GADGET- 3 is a hybrid code combining a tree algorithm (short-range force) with a high resolution particle-mesh (long-range force), RAMSES and ECOSMOG employ a multigrid relaxation method to solve the discretised Poisson equation on an adaptively-refined mesh. N-body codes like GADGET and ECOSMOG are imperative to follow structure formation well into the non-linear regime, where collapsed structures form. As we will explain in Chap. 5, this is particularly important in the case of modified gravity theories, where a vast amount of complex phenomenology is embedded in the non-linear equations of motion governing these theories.

All simulations presented in this thesis follow the evolution of the dark matter component only. Thus, the only relevant interaction between any pair of particles is the gravitational force between them. Dark matter-only simulations produce catalogues of haloes and subhaloes and it is the properties of these structures and statistics of the underlying density field that we wish to compare between different cosmological models.

1.3.3   Semi-analytic Galaxy Formation

The real Universe is, of course, made of more than just dark matter. To be able to test our cosmological simulations against observations, one needs to populate dark matter haloes with galaxies whose properties could be dependent on the properties and assembly history of the halo they are hosted in. One such approach, known as *semi-analytic* modelling, has become a particularly useful tool for better understanding the complex processes involved in galaxy formation and the connection between galaxies and their host haloes. Once a dark matter-only simulation has been run, a merger tree, which encapsulates the merging and accretion history of the dark matter haloes, can be constructed. A semi-analytic model (SAM) follows the properties of haloes in the merger tree and populates them with galaxies by solving a set of coupled differential equations treating the cooling of gas in haloes, star formation, feedback in the form of supernovae and AGN, chemical enrichment of the intergalactic medium, as well as the synthesis of stellar populations (White and Frenk 1991; Cole et al. 2000; Bower et al. 2006; Croton et al. 2006; Guo et al. 2011; Lacey et al. 2016). SAMs are typically characterised by a number of free parameters that are calibrated by requiring that the model reproduces a small selection of properties of the local galaxy population. In this thesis, the specific SAM we make use of is GALFORM (Cole et al. 2000; Lacey et al. 2016); the model is described in more detail in Sects. 3.3 and 4.2.

Semi-analytic modelling is not without its limitations. For example, unlike hydrodynamical simulations, a SAM cannot trace the flow of gas in and out of galaxies, nor can it accurately predict the response of dark matter halo properties to the presence of baryons or feedback processes. Furthermore, the essence of SAMs is that all equations pertaining to galaxy formation are dependent on the properties of the dark matter halo hosting the galaxy, which may only be a very crude approximation in some cases. That being said, however, SAMs possess one great advantage over hydrodynamical simulations, which is that they are computationally relatively inexpensive to run. This makes them ideal for rapidly exploring a vast parameter space – in terms of both the galaxy formation model itself, as well as the range of cosmological models being investigated. We exploit this fact in Chap. 4, where we apply GALFORM to a variety of sterile neutrino candidates.

References

Aarseth SJ (1963) MNRAS 126:223. https://doi.org/10.1093/mnras/126.3.223, http://adsabs.harvard.edu/abs/1963MNRAS.126.223A

Abazajian KN (2011) JCAP 3:010. https://doi.org/10.1088/1475-7516/2011/03/010, http://adsabs.harvard.edu/abs/2011JCAP...03.010A

Anderson ME, Churazov E, Bregman JN (2015) MNRAS 452:3905. https://doi.org/10.1093/mnras/stv1559, http://adsabs.harvard.edu/abs/2015MNRAS.452.3905A

Asaka T, Shaposhnikov M (2005) Phys Lett B 620:17. https://doi.org/10.1016/j.physletb.2005.06.020, http://adsabs.harvard.edu/abs/2005PhLB.620...17A

Babcock HW (1939) Lick Obs Bull 19:41. https://doi.org/10.5479/ADS/bib/1939LicOB.19.41B, http://adsabs.harvard.edu/abs/1939LicOB.19...41B

Barrow JD, Tipler FJ (1986) The anthropic cosmological principle

Bower RG, Benson AJ, Malbon R, Helly JC, Frenk CS, Baugh CM, Cole S, Lacey CG (2006) MNRAS 370:645. https://doi.org/10.1111/j.1365-2966.2006.10519.x, http://adsabs.harvard.edu/abs/2006MNRAS.370.645B

Boyarsky A, Ruchayskiy O, Iakubovskyi D, Franse J (2014) Phys Rev Lett 113:251301. https://doi.org/10.1103/PhysRevLett.113.251301, http://adsabs.harvard.edu/abs/2014PhRvL.113y1301B

Boyarsky A, Franse J, Iakubovskyi D, Ruchayskiy O (2015) Phys Rev Lett 115:161301. https://doi.
    org/10.1103/PhysRevLett.115.161301, http://adsabs.harvard.edu/abs/2015PhRvL.115p1301B
Boylan-Kolchin M, Bullock JS, Kaplinghat M (2011) MNRAS 415:L40. https://doi.org/10.1111/
    j.1745-3933.2011.01074.x, http://adsabs.harvard.edu/abs/2011MNRAS.415L.40B
Brooks AM, Zolotov A (2014) ApJ 786:87. https://doi.org/10.1088/0004-637X/786/2/87, http://
    adsabs.harvard.edu/abs/2014ApJ...786...87B
Bulbul E, Markevitch M, Foster A, Smith RK, Loewenstein M, Randall SW (2014) ApJ
    789:13. https://doi.org/10.1088/0004-637X/789/1/13, http://adsabs.harvard.edu/abs/2014ApJ...
    789...13B
Cappelluti N et al (2017). arXiv:1701.07932
Cholis I, Evoli C, Calore F, Linden T, Weniger C, Hooper D (2015) JCAP 12:005. https://doi.org/
    10.1088/1475-7516/2015/12/005, http://adsabs.harvard.edu/abs/2015JCAP...12.005C
Cole S, Lacey CG, Baugh CM, Frenk CS (2000) MNRAS 319:168. https://doi.org/10.1046/j.1365-
    8711.2000.03879.x, http://adsabs.harvard.edu/abs/2000MNRAS.319.168C
Collaboration Planck et al (2014) A&A 571:A16. https://doi.org/10.1051/0004-6361/201321591,
    http://adsabs.harvard.edu/abs/2014A%26A...571A.16P
Croton DJ et al (2006) MNRAS 365:11. https://doi.org/10.1111/j.1365-2966.2005.09675.x, http://
    adsabs.harvard.edu/abs/2006MNRAS.365...11C
Davis M, Efstathiou G, Frenk CS, White SDM (1985) ApJ 292:371. https://doi.org/10.1086/163168,
    http://adsabs.harvard.edu/abs/1985ApJ...292.371D
de Blok WJG, McGaugh SS, Bosma A, Rubin VC (2001) ApJL 552:L23. https://doi.org/10.1086/
    320262, http://adsabs.harvard.edu/abs/2001ApJ...552L.23D
Dodelson S, Widrow LM (1994) Phys Rev Lett 72:17. https://doi.org/10.1103/PhysRevLett.72.17,
    http://adsabs.harvard.edu/abs/1994PhRvL.72...17D
Fabian AC, Thomas PA, White RE III, Fall SM (1986) MNRAS 221:1049. https://doi.org/10.1093/
    mnras/221.4.1049, http://adsabs.harvard.edu/abs/1986MNRAS.221.1049F
Forman W, Schwarz J, Jones C, Liller W, Fabian AC (1979) ApJL 234:L27. https://doi.org/10.
    1086/183103, http://adsabs.harvard.edu/abs/1979ApJ...234L.27F
Governato F et al (2012) MNRAS 422:1231. https://doi.org/10.1111/j.1365-2966.2012.20696.x,
    http://adsabs.harvard.edu/abs/2012MNRAS.422.1231G
Guo Q et al (2011) MNRAS 413:101. https://doi.org/10.1111/j.1365-2966.2010.18114.x, http://
    adsabs.harvard.edu/abs/2011MNRAS.413.101G
Hooper D, Goodenough L (2011) Phys Lett B 697:412. https://doi.org/10.1016/j.physletb.2011.02.
    029, http://adsabs.harvard.edu/abs/2011PhLB.697.412H
Hubble E (1929) Proc Natl Acad Sci 15:168. https://doi.org/10.1073/pnas.15.3.168, http://adsabs.
    harvard.edu/abs/1929PNAS...15.168H
Jeltema T, Profumo S (2015) MNRAS 450:2143. https://doi.org/10.1093/mnras/stv768, http://
    adsabs.harvard.edu/abs/2015MNRAS.450.2143J
Jeltema T, Profumo S (2016) MNRAS 458:3592. https://doi.org/10.1093/mnras/stw578, http://
    adsabs.harvard.edu/abs/2016MNRAS.458.3592J
Katz N, White SDM (1993) ApJ 412:455. https://doi.org/10.1086/172935, http://adsabs.harvard.
    edu/abs/1993ApJ...412.455K
Kauffmann G, White SDM, Guiderdoni B (1993) MNRAS 264:201. https://doi.org/10.1093/mnras/
    264.1.201, http://adsabs.harvard.edu/abs/1993MNRAS.264.201K
Klypin A, Kravtsov AV, Valenzuela O, Prada F (1999) ApJ 522:82. https://doi.org/10.1086/307643,
    http://adsabs.harvard.edu/abs/1999ApJ...522...82K
Lacey CG et al (2016) MNRAS 462:3854. https://doi.org/10.1093/mnras/stw1888, http://adsabs.
    harvard.edu/abs/2016MNRAS.462.3854L
Laureijs R et al (2011). arXiv:1110.3193
Levi M et al (2013). arXiv:1308.0847
Li B, Zhao G-B, Teyssier R, Koyama K (2012) JCAP 1:051. https://doi.org/10.1088/1475-7516/
    2012/01/051, http://adsabs.harvard.edu/abs/2012JCAP...01.051L

Malyshev D, Neronov A, Eckert D (2014) Phys Rev D 90:103506. https://doi.org/10.1103/PhysRevD.90.103506, http://adsabs.harvard.edu/abs/2014PhRvD.90j3506M

Moore B, Ghigna S, Governato F, Lake G, Quinn T, Stadel J, Tozzi P (1999) ApJL 524:L19. https://doi.org/10.1086/312287, http://adsabs.harvard.edu/abs/1999ApJ...524L.19M

Oñorbe J, Boylan-Kolchin M, Bullock JS, Hopkins PF, Kereš D, Faucher-Giguère C-A, Quataert E, Murray N (2015) MNRAS 454:2092. https://doi.org/10.1093/mnras/stv2072, http://adsabs.harvard.edu/abs/2015MNRAS.454.2092O

Peebles PJE (1970) AJ 75:13. https://doi.org/10.1086/110933, http://adsabs.harvard.edu/abs/1970AJ.....75...13P

Peebles PJE (1982) ApJL 263:L1. https://doi.org/10.1086/183911, http://adsabs.harvard.edu/abs/1982ApJ...263L...1P

Perlmutter S et al (1999) ApJ 517:565. https://doi.org/10.1086/307221, http://adsabs.harvard.edu/abs/1999ApJ...517.565P

Potter D, Stadel J, Teyssier R (2016). arXiv:1609.08621

Riemer-Sørensen S (2016) A&A 590:A71. https://doi.org/10.1051/0004-6361/201527278, http://adsabs.harvard.edu/abs/2016A%26A...590A.71R

Riess AG et al (1998) AJ 116:1009. https://doi.org/10.1086/300499, http://adsabs.harvard.edu/abs/1998AJ....116.1009R

Rubin VC, Ford WK Jr (1970) ApJ 159:379. https://doi.org/10.1086/150317, http://adsabs.harvard.edu/abs/1970ApJ...159.379R

Sawala T et al (2016) MNRAS 457:1931. https://doi.org/10.1093/mnras/stw145, http://adsabs.harvard.edu/abs/2016MNRAS.457.1931S

Slipher VM (1915) Pop Astron 23:21. http://adsabs.harvard.edu/abs/1915PA.....23...21S

Smoot GF et al (1992) ApJL 396:L1. https://doi.org/10.1086/186504, http://adsabs.harvard.edu/abs/1992ApJ...396L...1S

Spergel DN et al (2003) APJS 148:175. https://doi.org/10.1086/377226, http://adsabs.harvard.edu/abs/2003ApJS.148.175S

Springel V (2005) MNRAS 364:1105. https://doi.org/10.1111/j.1365-2966.2005.09655.x, http://adsabs.harvard.edu/abs/2005MNRAS.364.1105S

Springel V et al (2008) MNRAS 391:1685. https://doi.org/10.1111/j.1365-2966.2008.14066.x, http://adsabs.harvard.edu/abs/2008MNRAS.391.1685S

Taylor AN, Dye S, Broadhurst TJ, Benítez N, van Kampen E (1998) ApJ 501:539. https://doi.org/10.1086/305827, http://adsabs.harvard.edu/abs/1998ApJ...501.539T

Teyssier R (2002) A&A 385:337. https://doi.org/10.1051/0004-6361:20011817, http://adsabs.harvard.edu/abs/2002A%26A...385.337T

Tyson JA, Valdes F, Wenk RA (1990) ApJL 349:L1. https://doi.org/10.1086/185636, http://adsabs.harvard.edu/abs/1990ApJ...349L...1T

Urban O, Werner N, Allen SW, Simionescu A, Kaastra JS, Strigari LE (2015) MNRAS 451:2447. https://doi.org/10.1093/mnras/stv1142, http://adsabs.harvard.edu/abs/2015MNRAS.451.2447U

von Hoerner S (1960) Zs f Astrophys 50

Weinberg S (1987) Phys Rev Lett 59:2607. https://doi.org/10.1103/PhysRevLett.59.2607, http://adsabs.harvard.edu/abs/1987PhRvL.59.2607W

White SDM, Frenk CS (1991) ApJ 379:52. https://doi.org/10.1086/170483, http://adsabs.harvard.edu/abs/1991ApJ...379...52W

White SDM, Frenk CS, Davis M (1983) ApJL 274:L1. https://doi.org/10.1086/184139, http://adsabs.harvard.edu/abs/1983ApJ...274L...1W

Zel'dovich YB (1970) A&A 5:84. http://adsabs.harvard.edu/abs/1970A%26A.....5...84Z

Zwicky F (1933) Helv Phys Acta 6:110. http://adsabs.harvard.edu/abs/1933AcHPh...6..110Z

Part I
Cosmology with Sterile Neutrinos

Chapter 2
Statistical Properties of Warm Dark Matter Haloes

2.1 Introduction

The identity of dark matter, the dominant matter component of the Universe, has long been a subject of great interest in cosmology.[1] In the last three decades, the model of non-relativistic dark matter consisting of heavy weakly-interacting particles with negligible thermal velocities at early times, the Cold Dark Matter (CDM) model, has become the cornerstone of the standard cosmological paradigm. The standard model with dark energy in the form of a cosmological constant, Λ (ΛCDM, henceforth just CDM) has been very successful in predicting and matching observational data on a wide range of scales, from the temperature fluctuations in the Cosmic Microwave Background (Planck Collaboration et al. 2014) to the statistics of galaxy clustering (Colless et al. 2001; Zehavi et al. 2002; Hawkins et al. 2003; Tegmark et al. 2004; Cole et al. 2005; Eisenstein et al. 2005; for a comprehensive review on the subject, see Frenk and White 2012).

With the advent of the LHC it was hoped that one of the best-motivated CDM candidates, the lightest supersymmetric particle (the neutralino) would be found. The lack of evidence for supersymmetry at the LHC and the absence of a convincing direct or indirect signal for CDM (but see Hooper and Goodenough 2011) has encouraged the exploration of viable alternatives. One of the most promising alternatives is the sterile neutrino (Dodelson and Widrow 1994; Asaka and Shaposhnikov 2005), which behaves as warm dark matter (WDM) due to the particles' non-negligible thermal velocities at early times. Being collisionless, this leads to free streaming and the damping of perturbations in the density field, creating a cutoff in the matter power spectrum on the scale of dwarf galaxies.

A simple extension of the Standard Model of particle physics, called the neutrino Minimal Standard Model (νMSM, Boyarsky et al. 2009), consists of three right-

---

[1] The content of this chapter is based on the article Bose et al. 'The Copernicus Complexio: statistical properties of warm dark matter haloes', Monthly Notices of the Royal Astronomical Society, Volume 455, Issue 1, pp. 318–333, published 30 October 2015. Reproduced with permission. All rights reserved, https://doi.org/10.1093/mnras/stv2294.

© Springer International Publishing AG, part of Springer Nature 2018
S. Bose, *Beyond ΛCDM* , Springer Theses,
https://doi.org/10.1007/978-3-319-96761-5_2

handed sterile neutrinos in which, for a specific choice of parameters, one of the sterile neutrinos behaves as a dark matter particle and the model explains neutrino flavour oscillations. Each one of this triplet of particles has its mass below the electroweak scale; one in the keV scale (denoted by M_1), and two in the GeV scale (denoted by M_2 and M_3). The former behaves as a relativistic particle at the time of neutrino decoupling and acts as WDM, and is then redshifted to non-relativistic energies during the radiation-dominated era. Unlike a thermal relic, the cutoff in the power spectrum introduced by a sterile neutrino of a fixed mass depends on a second parameter, the lepton asymmetry. As we explain later in the following section, it is possible to approximate the sterile neutrino power spectrum with a WDM thermal relic equivalent, particularly for very low and very high values of the lepton asymmetry.

The unidentified 3.53 keV X-ray line originally detected in the spectrum of a stack of galaxy clusters (Bulbul et al. 2014b) and in the spectra of M31 and the Perseus cluster (Boyarsky et al. 2014) could be a decay signal of sterile neutrino dark matter, with a particle mass of 7 keV. More recently, Boyarsky et al. (2015) have also identified a similar line in the centre of the Milky Way. While the excess at 3.5 keV has been seen in other studies (e.g. Urban et al. 2015), several groups have questioned the interpretation of this detection. For example, Riemer-Sørensen (2016) failed to find a signal in Chandra observations of the Milky Way. Of course, the Galactic centre is heavily contaminated by X-rays, which introduces uncertainties, a point made by Boyarsky et al. (2015).

Systematic effects can result from the atomic data used in modelling the plasma, as argued by Jeltema and Profumo (2015), who found no excess when re-analysing the Boyarsky et al. (2014) data and claimed that any signal at 3.5 keV could be explained by known Potassium (K XVIII) and Chlorine (Cl XVII) lines. Bulbul et al. (2014a) put this latter result down to the use of "incorrect atomic data and inconsistent spectroscopic modelling" by Jeltema and Profumo (2015). A further non-detection was then reported in the stacked spectra of galaxies from Chandra and XMM-Newton (Anderson et al. 2015), while most recently, Malyshev et al. (2014) analysed the spectra of stacked dwarf galaxies from XMM-Newton and claimed to rule out the Andromeda signal detected by Bulbul et al. (2014b) at the 4.6σ confidence level. This has spurred other groups (see, for example, Conlon and Day 2014) to associate the 3.53 keV signals with the conversion of a sterile neutrino into an axion, and its subsequent decay into photons. Such a mechanism requires a magnetic field, the presence and strength of which can vary from galaxy to galaxy, a scenario that could explain why this line is only seen in some objects.

Clearly, whether or not the 3.53 keV line really does correspond to a sterile neutrino decay remains an open question. It is, therefore, important to investigate the predictions for the formation of cosmic structures in a model in which the dark matter consists of particles that could decay producing such a line. Constraints on such models can be set from the observed clustering of the Lyman-α forest at high redshift whose small-scale structure would be erased if the dark matter were warm. On these grounds, Viel et al. (2013) recently set a (current) lower limit of 3.3 keV for the mass of a dominant thermal warm dark matter particle.

Coincidentally, the power spectrum of a 3.3 keV thermal warm dark matter particle is well approximated by that of a 7 keV sterile neutrino for a lepton asymmetry of $L_6 = 8.66$. This corresponds to the smallest allowed value of the power spectrum cutoff length (i.e. to the "coldest" power spectrum possible) for a sterile neutrino of mass 7 keV. This is the model that we will explore in this work. Ruling out this model from astronomical data on small scales would rule out the entire family of 7 keV sterile neutrino candidates. To investigate the model we use high resolution N-body simulations whose results we compare with those of CDM simulations with the same phases in the initial conditions. We are interested exclusively in characterising the properties of dark matter haloes of mass in the region of the power spectrum cutoff and, in this study, we ignore the effects of baryons. Such effects must be taken into account when comparing model predictions with observations. In the case of CDM, relevant baryon effects on the small scales of interest here have recently been quantified by Sawala et al. (2013), (2015), (2016a), Schaller et al. (2015)

The layout of this chapter is as follows. In Sect. 2.2 we introduce the concept of sterile neutrinos, and some terminology that will be important for the rest of this chapter, as well as Chaps. 3 and 4. In Sect. 2.3, we introduce the simulations used in this work, the modelling of the WDM component, and describe how we tackle the issue of spurious halo formation in our simulations. In Sect. 2.4 we present our main results from the comparison of WDM and CDM from our simulations, in terms of both the large-scale distribution of matter, and the internal structure of haloes. Finally, in Sect. 2.5, we summarise our findings and look into some future work that will be carried out with the same set of simulations.

2.2 The Sterile Neutrino Model

Sterile neutrinos[2] are relativistic when they decouple and therefore have non-negligible velocities which smear out density perturbations on small scales. Hence, sterile neutrinos behave as WDM. In the original model introduced by Dodelson and Widrow (1994), sterile neutrinos are created by non-resonant mixing with active neutrinos in the Standard Model. The scale of the free streaming is determined solely by the rest mass of the sterile neutrino – the lighter the particle, the larger the free streaming length, and the larger the scales at which differences relative to CDM appear.

Shi and Fuller (1999) proposed an alternative production mechanism in which the abundance of sterile neutrinos is boosted by a primordial *lepton asymmetry*. The value of this quantity, which measures the excess of leptons over anti-leptons, affects the scale of free streaming in addition to the rest mass of the sterile neutrino. Asaka and Shaposhnikov (2005) proposed a model for the generation of the lepton asymmetry by introducing three right-handed sterile neutrinos in what is known as

---

[2]These particles are 'sterile' in the sense that they do not interact via the weak force, as is the case for active neutrinos in the Standard Model.

the 'Neutrino Minimal Standard Model' (νMSM, see also Boyarsky et al. 2009). In this model, a keV mass sterile neutrino (labelled N_1) is partnered with two GeV mass sterile neutrinos (N_2 and N_3). It is N_1 that behaves as the dark matter, with its keV mass (M_1) leading to early free streaming. The decay of N_2 and N_3 prior to the production of N_1 generates significant lepton asymmetry; this boosts the production of N_1 via resonant mixing. Here, we formally quantify the lepton asymmetry, or L_6, as:

$$L_6 \equiv 10^6 \left(\frac{n_{\nu_e} - n_{\bar{\nu}_e}}{s} \right), \qquad (2.2.1)$$

where n_{ν_e} is the number density of electron neutrinos, $n_{\bar{\nu}_e}$ the number density of electron anti-neutrinos and s is the entropy density of the Universe (Laine and Shaposhnikov 2008). The scales at which the power spectrum is suppressed for sterile neutrinos vary non-monotonically as a function of L_6. If L_6 is very small ($\ll 1$) the power spectrum exhibits a similar abrupt cutoff to that of a thermal relic. As L_6 is increased, the cutoff becomes gentler and k_{hm} shifts to larger values. At some value of L_6 (typically between 8 and 25 depending on the sterile neutrino mass), k_{hm} reaches a maximum; for still higher L_6, k_{hm} retreats to lower k and returns to its original shape and position (Shi and Fuller 1999; Abazajian 2014).

A third parameter in the νMSM is the *mixing angle*, θ_1. The requirement that the model should achieve the correct dark matter abundance for a given sterile neutrino rest mass uniquely fixes the value of θ_1 for a particular choice of L_6. The X-ray flux, F, associated with the decay of N_1 is then proportional to $\sin^2 (2\theta_1) M_1^5$. We refer the reader to Venumadhav et al. (2016) and Lovell et al. (2016) for a more comprehensive discussion of the sterile neutrino model.

2.3 The Copernicus Complexio Simulations

In this section, we provide an overview of the initial conditions and modelling of the WDM component in our simulations.

2.3.1 The Simulation Set-Up

The N-body simulations presented in this chapter are part of the *COpernicus COmplexio* (COCO) simulation programme (Hellwing et al. 2016a) being carried out by the *Virgo Consortium*. This is a set of cosmological "zoom-in" simulations (Katz and White 1993; Frenk et al. 1996), as was done in the GIMIC simulations (Crain et al. 2009). The parent simulation, called the *COpernicus complexio LOw Resolution* (or COLOR) simulation, followed the evolution of 4.25 billion particles in a periodic box of size 70.4 h^{-1} Mpc. We extracted a roughly spherical region of radius $\sim 18h^{-1}$ Mpc, and centred on the location $(42.2, 51.2, 8.8)$ h^{-1} Mpc in the COLOR volume. Both COLOR and COCO assume cosmological parameters derived from the seven-year *Wilkinson Microwave Anisotropy Probe* (WMAP 7) data (Komatsu et al. 2011), with the parameters: $\Omega_m = 0.272, \Omega_\Lambda = 0.728, h = 0.704, n_s = 0.967$ and $\sigma_8 = 0.81$. Here, $\Omega_{\{m,\Lambda\}}$ represents the present-day fractional contribution of matter and the cosmological constant respectively, in units of the critical density $\rho_c = 3H_0^2/8\pi G$, $h = H_0/100$ km/s/Mpc is the dimensionless Hubble parameter, n_s is the spectral index of the primordial power spectrum, and σ_8 is the linear *rms* density fluctuation in a sphere of radius $8\,h^{-1}$ Mpc at $z = 0$.

Dark matter particles with three different masses are used in regions simulated at different resolutions within the parent simulation volume. Initially, the high-resolution region has a shape similar to an amoeba which approximates a sphere of radius $\sim 17.4\,h^{-1}$ Mpc at the present time. It contains 12.9 billion particles of mass $1.135 \times 10^5\,h^{-1}\,M_\odot$. The volume surrounding this region contains the medium- $(3.07 \times 10^6\,h^{-1}\,M_\odot)$ and low-resolution $(1.96 \times 10^8\,h^{-1}\,M_\odot)$ particles. We have taken care to minimise contamination of the high-resolution region by lower mass particles and all the haloes discussed in this study are entirely made up of the high-resolution particles. The gravitational softening was kept fixed at $\varepsilon \sim 230\,h^{-1}$ pc for the high-resolution particles, increasing by a factor of 10 each time for the medium- and low-resolution particles.

The simulation ran from $z = 127$ to $z = 0$ using the GADGET- 3 code, which is an updated version of the publicly available GADGET- 2 code (Springel et al. 2001a; Springel 2005). Phase information for the creation of the initial conditions for both COCO- WARM and COCO- COLD was obtained from the public Gaussian white noise field PANPHASIA (Jenkins 2013), and perturbations thereafter were calculated using the second-order Lagrangian Perturbation Theory (2LPT) algorithm presented in Jenkins (2010). The details of the simulation, along with the PANPHASIA phase descriptor, are summarised in Table 2.1.

The distinctive feature of WDM particles are non-negligible thermal velocities at early times, which result in free streaming that washes out perturbations in the matter distribution below the free streaming scale (Bond and Szalay 1983; Schneider et al. 2012; Benson et al. 2013). As a result, we expect the abundance, distribution and internal structure of WDM haloes to be different from those of CDM haloes. Indeed, thermal velocities introduce a limit to the fine-grained phase space density in dark matter haloes, creating cores in the density profile (Macciò et al. 2012; Shao et al. 2013). However, as shown in these papers, the cores produced by realistic thermal relics are only a few parsecs in size, and thus not astrophysically relevant. In our simulations we can neglect these thermal velocities, which at $z = 0$ are of the order of a few tens of metres per second (Lovell et al. 2012) so, over the course of the simulation, which starts at $z = 127$, the particles would travel only a few kiloparsecs, comparable to the mean interparticle spacing of the high-resolution particles.

The WDM power spectrum of density fluctuations is often modelled by the transfer function, $T(k)$, relative to the CDM case:

$$P_{\mathrm{WDM}}(k) = T^2(k) P_{\mathrm{CDM}}(k) . \qquad (2.3.1)$$

Table 2.1 Cosmological parameters used in the COCO simulations, and its parent volume, COLOR. Here, m_{WDM} is the mass of the thermal relic warm dark matter particle, N_p is the total number of particles (of all types) used in the simulation, V_{hr} is the approximate volume of the high-resolution region at $z = 0$, $m_{p,hr}$ is the mass of an individual high-resolution dark matter particle, $N_{p,hr}$ is the total number of particles of this species, whereas ε_{hr} is the softening length applied to them. The cosmological parameters h, Ω_m, Ω_Λ and σ_8 are as described in the text. The phases for the parent COLOR simulation can be generated using the PANPHASIA phase descriptor provided in the last row. The blank fields in the COCO column mean that the parameter assumes the same value as in the parent simulation, COLOR

Parameter	COLOR (Parent volume)	COCO (This chapter)
Box Size	$70.4\,h^{-1}$ Mpc	–
m_{WDM}	3.3 keV	–
N_p	4, 251, 528, 000	13, 384, 245, 248
V_{hr}	$70.4^3\,h^{-3}\mathrm{Mpc}^3$	$\sim 2.2 \times 10^4\,h^{-3}\mathrm{Mpc}^3$
$m_{p,hr}$	$6.196 \times 10^6\,h^{-1}\,M_\odot$	$1.135 \times 10^5\,h^{-1}\,M_\odot$
$N_{p,hr}$	4, 251, 528, 000	12, 876, 807, 168
ε_{hr}	$230\,h^{-1}$ pc	–
h	0.704	–
Ω_m	0.272	–
Ω_Λ	0.728	–
σ_8	0.81	–
Phase descriptor	[Panph1,L16,(31250,23438,39063), S12,CH582187950,COLOR]	–

We approximate $T(k)$ using the fitting formula provided by Bode et al. (2001):

$$T(k) = \left(1 + (\alpha k)^{2\nu}\right)^{-5/\nu} , \qquad (2.3.2)$$

where α and ν are constants. As computed by Viel et al. (2005), for $k < 5\,h^{-1}$ Mpc, the value $\nu = 1.12$ provides the best-fitting transfer function. The value of α is dependent on the mass of the WDM particle (Viel et al. 2005):

$$\alpha = 0.049 \left[\frac{m_{WDM}}{\mathrm{keV}}\right]^{-1.11} \left[\frac{\Omega_{WDM}}{0.25}\right]^{0.11} \left[\frac{h}{0.7}\right] h^{-1}\ \mathrm{Mpc}, \qquad (2.3.3)$$

and determines the scale of the cutoff due to free streaming in the WDM power spectrum relative to CDM. It should be noted that this transfer function is a fit to the full thermal relic power spectrum, obtained by solving the Boltzmann equation.

As we can see in Eq. 2.3.3, the "warmer" the dark matter particle (i.e., the lower its rest mass is), the larger the scale at which the cutoff in the power spectrum occurs.

One way to define the characteristic scale in the power spectrum is through the "half-mode" wavenumber, k_{hm}, where the transfer function in Eq. 2.3.2 drops by a factor of two:

$$k_{\mathrm{hm}} = \frac{1}{\alpha} \left(2^{\nu/5} - 1 \right)^{1/2\nu} . \tag{2.3.4}$$

The associated "half-mode mass", M_{hm}, is the mean density enclosed within this half-mode:

$$M_{\mathrm{hm}} = \frac{4\pi}{3} \bar{\rho} \left(\frac{\lambda_{\mathrm{hm}}}{2} \right)^3 . \tag{2.3.5}$$

For the 3.3 keV model, this occurs at around $M_{\mathrm{hm}} \sim 2 \times 10^8\ h^{-1} M_\odot$ (Colín et al. 2008; Angulo et al. 2013; Viel et al. 2013). We will show later that differences in the formation time of haloes in WDM and CDM begin to appear below $\sim 2 \times 10^9\ h^{-1} M_\odot$, approximately an order of magnitude above the half-mode mass scale.

The power spectra used in the COCO simulations are shown as thick lines in Fig. 2.1: CDM in black, 3.3 keV WDM in red and 7 keV sterile neutrinos with $L_6 = 8.66$ in blue. All three power spectra agree on large scales. On small scales, the two warm dark matter models differ from CDM. k_{hm} for the sterile neutrino case occurs at a very similar scale, and the cutoff has a similar shape to that for the thermal relic case. On smaller scales still, the sterile neutrino power spectrum has more power than its thermal counterpart, but the differences only become significant on scales where the

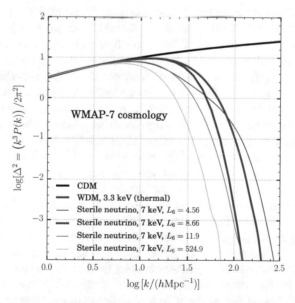

Fig. 2.1 The (dimensionless) matter power spectrum for: a thermal 3.3 keV WDM (red), a sterile neutrino of mass $m_{\nu_s} = 7$ keV and lepton asymmetry $L_6 = 8.66$ (blue) and CDM (black). Both the WDM and sterile neutrino power spectra have significantly suppressed power at small scales, with the deviation from CDM case at almost identical scales: $\log(k) \gtrsim 1.0\,h\,\mathrm{Mpc}^{-1}$. Also shown as thin coloured lines are power spectra for 7 keV sterile neutrinos with different values of L_6, as indicated in the legend. Figure reproduced from Bose et al. (2016)

amplitude is, at most, a few percent of the peak amplitude. These differences are negligible and can be safely ignored in our simulations. The thin lines in the figure correspond to 7 keV sterile neutrino power spectra for different values of the lepton asymmetry, L_6. The $L_6 = 8.66$ model that we have simulated corresponds to the "coldest" possible 7 keV sterile neutrino.

The COCO simulations are amongst the highest resolution WDM N-body simulations of a cosmological volume performed to date. Previous simulations at higher mass resolution have focussed on properties of individual haloes (e.g. Lovell et al. 2014, Colín et al. 2015). Other WDM simulations of comparable mass resolution to ours (e.g. Schneider et al. 2013) followed smaller volumes. The advantage of the relatively high mass resolution and large volume of COCO is that it provides a large statistical sample of well-resolved dark matter haloes spanning nearly seven decades in mass. In particular, resolving the halo mass function down to $\sim 10^7 - 10^8 \, h^{-1} \, M_\odot$, as COCO does, is a crucial input to studies that attempt to distinguish amongst different types of dark matter using strong gravitational lensing (Vegetti and Koopmans 2009; Li et al. 2016).

2.3.2 Halo Identification and Matching

Haloes were identified in our simulations using the friend-of-friend (FOF) algorithm (Davis et al. 1985) with a linking length of 0.2 times the mean interparticle separation, and a minimum of 20 particles. Gravitationally-bound substructures within these groups were then identified using the SUBFIND algorithm (Springel et al. 2001b), although in this chapter, we will be mostly concerned with the properties of the WDM FOF groups. We determine the halo centre using the "shrinking sphere" method of Power et al. (2003). In short, we recursively compute the centre of mass of all particles within a shrinking sphere, until a convergence criterion is met. In each iteration, the radius of the sphere is reduced by 5%, and stopped when only 1000 particles or 1% of the particles of the initial sphere (whichever is smaller) are left.

Comparing halo statistics between sets of simulations requires consistent definitions for the various properties of the haloes. In this work, we make use of two definitions of mass: M_{FOF}, which is the mass of all particles identified by the algorithm as belonging to the FOF group, and M_{200}, which is the mass contained within a sphere of radius r_{200} (centred on the "shrinking sphere" centre defined above), within which the average density is 200 times the critical density of the Universe (ρ_c) at the specified redshift. Another common radius used to define a halo edge is the virial radius, r_{vir}, within which the density of the halo $\bar{\rho}(< r_{\mathrm{vir}}) = \Delta \rho_c$, where $\Delta \sim 178 \Omega_m^{0.45}$ (motivated by the spherical collapse model, Eke et al. 1996. Note that this definition is consistent with the virial overdensity relation in Bryan et al. 1998). Table 2.2 summarises the total number of groups and self-bound substructures identified at $z = 0$ in our simulations.

Since both COCO- WARM and its COLD counterpart were simulated using the same initial phases, we are able to match many objects between the two simulations. This

Table 2.2 Number of groups and subhaloes identified by the FOF algorithm and SUBFIND in COLOR and COCO at $z = 0$

Simulation	$N_{FOF}(z = 0)$	$N_{subs}(z = 0)$
COLOR- COLD	3, 961, 192	4, 770, 041
COLOR- WARM	2, 609, 122	3, 082, 275
COCO- COLD	8, 896, 811	10, 502, 187
COCO- WARM	2, 548, 743	2, 830, 514

also allows us to make object-by-object comparisons in addition to comparing just statistical distributions of halo properties. In order to correctly match the haloes we do the following: first, we take the 50 most-bound particles from a COCO- WARM halo, and look for the COCO- COLD halo in which there are at least 25 (50%) of these particles. We then confirm the match by repeating the same process, this time starting with the COCO- COLD haloes, in decreasing order of mass. This results in a bijective match between haloes in the two simulations. Using this method, we are able to match around 97% of haloes with $M_{200} > 10^8 \, h^{-1} \, M_\odot$.

2.3.3 Spurious Haloes and Their Removal

Number counts of haloes and subhaloes are fundamental statistics of the halo population, so the correct identification of haloes is of primary importance. It has been known for some time (Wang and White 2007; Angulo et al. 2013; Lovell et al. 2014) that in simulations in which the initial power spectrum has a resolved cutoff, as is the case for COCO- WARM, small-scale structure is seeded in part by the discreteness of the particle set. In other words, a substructure finder will identify density peaks that have arisen not as a result of gravitational instabilities from a cosmological perturbation, but rather due to gravitational instability from noise. These artificial fragments can often by identified "by eye" as they tend to be regularly spaced along filaments of the mass distribution. They produce a power-law-like upturn at small masses in the WDM mass function. Since this is just a numerical (and resolution-dependent) artefact of our WDM simulations, care must be taken to identify these spurious haloes and, if appropriate, remove them from the halo catalogue. While it is, in principle, possible to eliminate these structures by increasing the resolution of the simulation, this is computationally prohibitive: Wang and White (2007) have shown that the mass at which spurious structures dominate the mass function scales with the number of particles in the simulation, N, as $M \propto N^{-1/3}$.

Ideally, one would like to prevent the spurious fragmentation from occurring in the first place. Spurious halo formation, fundamentally, is a consequence of discretising the collisionless fluid of dark matter using particles. Hahn et al. (2013) found that a more accurate representation of the dark matter fluid that properly tracks folds

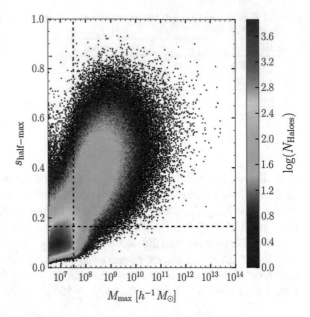

Fig. 2.2 Number density of haloes in the sphericity vs. maximum mass space in COCO- WARM at $z = 0$. The dashed black lines show the cuts on sphericity and mass that we use to clean the halo catalogue. Rejected (spurious) candidates are those that fail the cuts in the manner described in the text. Figure reproduced from Bose et al. (2016)

in the three dimensional phase-space sheet significantly reduced the formation of spurious fragments. Unfortunately, this approach is prohibitively expensive in regions of multiple foliations of the phase sheet, such as at the centres of dark matter haloes. Recognising that spurious fragmentation arises in regions of large anisotropic force errors, Hobbs et al. (2016) devised a scheme of anisotropic force softening, which also suppresses spurious halo formation. However, the numerical convergence and applicability of this technique over a wide range of cosmological scales remains an open question.

Lovell et al. (2014) developed an algorithm to identify spurious clumps in WDM simulations. A large number of them can be removed by performing a mass cut below a resolution-dependent scale, as suggested by Wang and White (2007):

$$M_{\text{lim}} = 10.1 \bar{\rho} \, d \, k_{\text{peak}}^{-2} \, , \tag{2.3.6}$$

where d is the mean interparticle separation and k_{peak} is the spatial frequency at which the dimensionless power spectrum, $\Delta^2(k)$, has its maximum. Applying this condition on its own would also remove some genuine haloes that form below this scale. Lovell et al. (2014) refined this criterion by also making a cut on the basis of the shapes of the initial Lagrangian regions from which WDM haloes form. They find that the spurious candidates tend to have much more flattened configurations in their (unperturbed) initial positions than genuine haloes, as judged from a CDM simulation. Defining the sphericity, s, of haloes as the axis ratio, c/a, of the minor to major axes in the diagonalised moment of inertia tensor of the initial particle load, the sphericity cut is made such that 99% of the CDM haloes at that redshift lie above the threshold.

Following exactly the methodology of Lovell et al. (2014), we clean the COCO-WARM catalogue as follows: (1) remove all (sub)haloes with $s_{half-max} < 0.165$,[3] irrespective of mass; (2) for those that pass (1), remove (sub)haloes with $M_{max} < 0.5M_{lim}$. Here, M_{max} is the maximum mass attained by a (sub)halo during its evolution, and $s_{half-max}$ is the sphericity ($= c/a$) of the (sub)halo at the half-maximum mass snapshot. This is chosen so as to identify a (sub)halo at a time well before it falls into a larger host, when its particles are subject to tidal stripping. The factor of 0.5 in condition (2) is calibrated by matching between resolutions in the AQUARIUS simulations (see Lovell et al. 2014 for details). Having done so, we find that over 91% of the (FOF) haloes formed in COCO-WARM are in fact spurious, and are rejected from the halo catalogue when computing properties like mass functions. The elements of this section are summarised in Fig. 2.2.

2.4 Results

In both cold and warm models, dark matter haloes assemble in a hierarchical way, acquiring mass by merging with other haloes and by smoothly accreting ambient mass (e.g. Press and Schechter 1974; Frenk et al. 1985; Lacey and Cole 1993; Wechsler et al. 2002). In this section, we focus on global halo properties such as formation times, abundance and internal structure. We make a direct comparison between our cold and warm dark matter models. On scales much larger than the WDM suppression scale in the initial power spectrum, we expect the properties of haloes to be very similar in the two cases, but differences should become increasingly important at $\sim 2 \times 10^9 \, h^{-1} \, M_\odot$ and below.

2.4.1 Redshift of Formation

The absence of primordial perturbations below the cutoff scale in the WDM power spectrum induces differences in the formation times of the smallest haloes. We can visualise these differences directly by examining the images displayed in Fig. 2.3. At early times, the projected density in COCO-WARM (right panels) is visibly smoother than the equivalent projection in COCO-COLD (left panels), which has a "grainier" appearance owing to the very large number of haloes below $\sim 10^9 \, h^{-1} \, M_\odot$ that form in this case, well before the first objects have collapsed in COCO-WARM. Thus, the onset of the structure formation process in this simulation is delayed relative to its CDM counterpart.

---

[3]The criterion $s_{half-max} < 0.165$ is appropriate for haloes identified at $z = 0$; for higher redshifts, one needs to determine the 1% sphericity cut at *that* redshift.

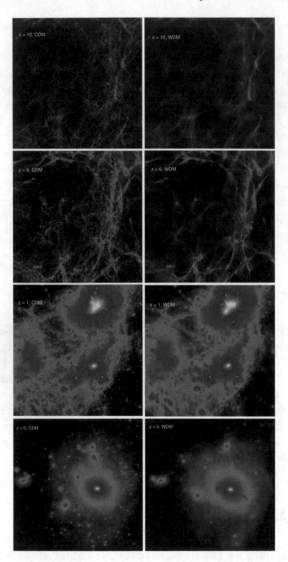

Fig. 2.3 Redshift evolution of the projected dark matter density in COCO- COLD (left) and the 3.3 keV COCO- WARM Universe (right). From top to bottom, the top three panels show snapshots at $z = 10$, $z = 6$, $z = 1$ of the projected mass density in cubes of side $2\,h^{-1}$ Mpc, centred on the most massive group at $z = 0$. The bottom panels show zooms of a $5 \times 10^{10}\,h^{-1}\,M_\odot$ halo at $z = 0$ in a cube of side $150\,h^{-1}$ kpc. The emergence of small haloes at early times is apparent in the CDM case, when the WDM distribution is much smoother. The formation of large haloes occurs at roughly the same time in the two simulations and the subsequent growth of these haloes is similar in the two cases. In the zoom shown in the bottom panel, the lack of substructure in the WDM case compared to its CDM counterpart is stark. Figure reproduced from Bose et al. (2016)

In order to quantify the different halo formation epochs in COCO- WARM and COCO- COLD, we trace the evolution of each FOF group through its merger tree, and define the redshift of formation as the first time when the mass of the most massive progenitor exceeds half the final FOF mass: $M(z_{form}) = M(z = 0)/2$ (e.g. Harker et al. 2006; Neto et al. 2007). Other definitions of halo formation time also exist in the literature (e.g. Navarro et al. 1996, 1997), which should be borne in mind when making comparisons.

The result, for *all* haloes in COCO- WARM (including spurious objects) and COCO- COLD is shown in Fig. 2.4. The formation redshifts of haloes of mass $M_{200} \gtrsim 2 \times 10^9 \ h^{-1} \ M_\odot$, are very similar in COCO- WARM and COCO- COLD, as expected. The difference between the two begins to manifest itself below a mass of $M_{200} \sim 2 \times 10^9 \ h^{-1} \ M_\odot$, an order of magnitude above the half-mode mass scale for a 3.3 keV WDM particle (c.f. Sect. 2.3.1). For these smaller haloes, z_{form} is lower for WDM than CDM. The sudden upturn in the WDM z_{form} for $M_{200} < 10^8 \ h^{-1} \ M_\odot$ (shown in the open red circles) is a signature of the spurious haloes described in Sect. 2.3.3. From here on, we will exclude these spurious haloes and only show results from the cleaned COCO- WARM sample. The difference in formation times is a subject we

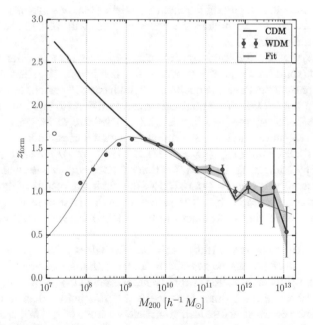

Fig. 2.4 The median redshift of formation of all FOF groups in COCO- WARM and COCO- COLD, as a function of the halo mass, M_{200}. The redshift z_{form} is defined in the text. The error bars/shaded region represent the bootstrapped errors on the median in each mass bin in COCO- WARM and COCO- COLD respectively. As expected, there is good agreement at the high-mass end, whereas the differences between CDM and WDM become apparent below $\sim 2 \times 10^9 \ h^{-1} \ M_\odot$. The thin red line is a fit to the COCO- WARM redshift of formation, using Eq. 2.4.1. Figure reproduced from Bose et al. (2016)

will revisit when comparing the concentration-mass relations of WDM and CDM in Sect. 2.4.4. Note that in this figure, we include all haloes, and not necessarily matched between CDM and WDM, which is why the medians at the largest mass bins are not exactly identical.

We find that the delay in the formation time of COCO- WARM haloes of a given mass, relative to COCO- COLD, is well described by the fitting function:

$$\frac{z_{\text{form}}^{\text{WDM}}}{z_{\text{form}}^{\text{CDM}}} = \left(1 + a\frac{M_{\text{hm}}}{M_{200}}\right)^{-b}, \qquad (2.4.1)$$

where M_{hm} is the half-mode mass introduced in Sect. 2.3.1, $a = 1.23$ and $b = 0.56$. This fit is shown as the thin red line in Fig. 2.4.

2.4.2 Differential Halo Mass Functions

Counting the number of dark matter haloes as a function of their mass is one of the simplest and most important population statistics that one can use to distinguish between WDM and CDM models, since fewer haloes will form in the former close to the half-mode mass.

In Fig. 2.5, we show the build-up of the halo population as a function of redshift in COCO- COLD (solid lines) and COCO- WARM (dashed lines). The shaded regions and error bars represent the Poisson uncertainty in both cases. Spurious haloes have been omitted from the WDM differential halo mass function (dHMF) at each redshift, using the methodology outlined in Sect. 2.3.3. The edge of the grey region marks the nominal resolution limit of our simulation which corresponds to a halo with at least 300 particles within r_{200} ($M_{200} \sim 3.4 \times 10^7$ h^{-1} $M_\odot$). This 300-particle limit was derived by comparing the mass function of COCO- COLD with that of its lower-resolution counterpart COLOR- COLD. Below this limit, the results of the simulations become increasingly unreliable. The results at high masses are noisy because of the small number of high-mass haloes formed in the relatively small volume of our simulations.

The general trend across redshifts is similar: for haloes with $M_{200} > 2 \times 10^9$ h^{-1} $M_\odot$, the dHMF in COCO- WARM and COCO- COLD are almost identical. The abundance of haloes below this mass scale is strongly suppressed in COCO- WARM, to the extent that, at $z = 10$, there are 5 times fewer $\sim 10^8$ h^{-1} $M_\odot$ haloes than in COCO- COLD. The delayed non-linear structure formation below $\sim 2 \times 10^9$ h^{-1} $M_\odot$ can also be seen from the fact that there are as many haloes with $M_{200} = 10^8$ h^{-1} $M_\odot$ in COCO- WARM at $z = 10$, as there are haloes with $M_{200} = 6 \times 10^8$ h^{-1} $M_\odot$ in COCO- COLD at that redshift.

Within the CDM paradigm, there are a number of analytic predictions for the differential halo mass function (dHMF), notably the Press-Schechter formula (Press and Schechter 1974; Bond et al. 1991; Lacey and Cole 1993), and the ellipsoidal

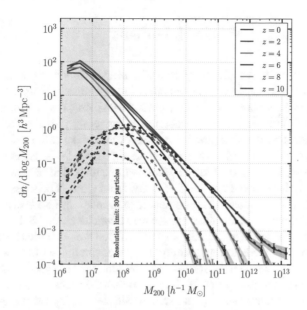

Fig. 2.5 The redshift evolution of the halo mass function in COCO- COLD and COCO- WARM. The solid lines show the CDM results, with the shaded regions representing the associated 1σ Poisson errors. The dashed lines with error bars represent the equivalent relation from COCO- WARM, with spurious haloes removed. The different colours show results for a selection of redshifts, as indicated in the legend. The grey shaded region corresponds to haloes with fewer than 300 particles within r_{200}. Figure reproduced from Bose et al. (2016)

collapse model (ST; Sheth and Tormen 1999, although this model is not fully analytic since it is tuned to numerical simulations). The dHMF is given by:

$$\frac{\mathrm{d}n}{\mathrm{d}\log M} = \frac{\bar{\rho}}{M} f(\nu) \left| \frac{\mathrm{d}\log \sigma^{-1}}{\mathrm{d}\log M} \right|, \qquad (2.4.2)$$

where $f(\nu)$ is the so-called *halo multiplicity function* and for hierarchical cosmologies has a universal form (see e.g. Jenkins et al. 2001; Reed et al. 2007; Tinker et al. 2008; Angulo et al. 2012). In the ST formalism, it is approximated by:

$$f(\nu) = A\sqrt{\frac{2q\nu}{\pi}} \left[1 + (q\nu)^{-p} \right] e^{-q\nu/2}. \qquad (2.4.3)$$

Here, $\nu \equiv \delta_c^2(z)/\sigma^2(M)$, $A = 0.3222$, $q = 0.707$ and $p = 0.3$. In linear theory, $\delta_c(z) \equiv 1.686/D(z)$, where $D(z)$ is the linear growth rate of perturbations. The value of δ_c is appropriate for the Einstein-de Sitter model, but differs slightly in ΛCDM due to a weak dependence on $\Omega_m(z)$. Finally, $\sigma^2(M)$ is the variance in the mass density field on mass scale, M, given by:

$$\sigma^2(M) = \int \frac{dk}{k} \Delta^2(k) \tilde{W}^2(k, M).$$ (2.4.4)

Here, $\tilde{W}(k, M)$ is the Fourier transform of a window function containing mass M, and $\Delta^2(k)$ is the dimensionless power spectrum as defined in Fig. 2.1.

In the Press-Schechter and Sheth-Tormen formalisms, the *rms* fluctuation amplitude, $\sigma^2(M)$, is assumed to be a monotonically increasing function of M. This is no longer true for the truncated power spectrum of WDM, so care must be taken when choosing an appropriate window function. In the CDM, $W(k, M)$ is usually chosen to be the real-space spherical top-hat function, a choice that results in an excellent match to the dHMF in cosmological N-body simulations. The same for WDM predicts an excess of low-mass haloes compared to simulations (Bode et al. 2001; Menci et al. 2012; Schneider et al. 2012, but see also Schneider et al. 2013). This problem was solved by Benson et al. (2013), who generalised the (extended) Press-Schechter formalism by using the correct solution for the excursion set barrier first-crossing distribution in WDM models. Rather than the top-hat real-space window function, they used a sharp k-space filter for WDM, so that the variance, $\sigma(M)$, remains flat up to the half-mode mass and then declines with increasing mass (see Fig. 2.6). In this formalism the smoothing scale, R, is defined as:

$$R = \frac{a}{k_s},$$ (2.4.5)

Fig. 2.6 The fractional variance of the density field, $\sigma(M)$, calculated in Eq. 2.4.4 using a top-hat filter in real-space for CDM, and a sharp k-space filter for WDM. The flattening of the relation below $10^8 \, h^{-1} \, M_\odot$ is due to the suppression of power below these scales in WDM, relative to CDM. The dashed line indicates the upper limit to the halo masses formed in our volume-limited simulations. Figure reproduced from Bose et al. (2016)

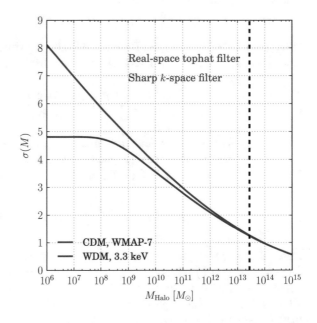

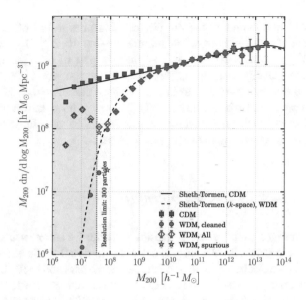

Fig. 2.7 Differential halo mass functions from the COCO- WARM and COCO- COLD simulations, compared to the predictions of the ellipsoidal collapse formalism of Sheth and Tormen (1999). The solid lines show the predictions of the standard formalism applied to CDM; the dashed lines show the predictions of the modified, sharp k-space filter of Benson et al. (2013). The symbols represent results from our simulations as denoted in the legend: blue squares for COCO- COLD, green diamonds for all COCO- WARM FOF haloes, red circles for the genuine haloes and yellow stars for spurious haloes. The grey shaded region corresponds to haloes with fewer than 300 particles within r_{200}. Figure reproduced from Bose et al. (2016)

where $k_s = 2\pi\kappa/\alpha$, α as defined in Eq. 2.3.3, $\kappa = 0.361$ and $a = 2.5$. Benson et al. (2013) choose the free parameters such that the theoretical mass function turns over at the same scale as the halo mass function from simulations. This choice of parameters should be applicable to all thermal WDM models, since the effect of the WDM suppression is captured in the value of α (Eq. 2.3.3).

In Fig. 2.7, we compare the $z = 0$ dHMF for COCO- COLD (blue squares), the full COCO- WARM (genuine and spurious objects; green diamonds), the spurious COCO- WARM objects only (yellow stars) and the genuine COCO- WARM haloes only (red circles).

The solid and dashed black lines in Fig. 2.7 show the ST predictions for the mass functions in CDM and WDM respectively. For $M_{200} > 2 \times 10^9\,h^{-1}\,M_\odot$, the mass functions for CDM and WDM trace one another exactly, as expected. Below this mass, the WDM mass function begins to peel off from the CDM case, reaching half the CDM amplitude at $M_{200} \sim 2 \times 10^8\,h^{-1}\,M_\odot$. This agrees with the half-mode mass scale, M_{hm}, introduced in Sect. 2.3.1. The raw WDM mass function (green diamonds) is entirely dominated by the spurious objects (yellow stars) below $\sim 4 \times 10^7\,h^{-1}\,M_\odot$, where the mass function shows an artificial upturn. On the other hand, the cleaned

WDM sample, represented by the red circles, continues to fall off smoothly from the regime free of artificial haloes. The feature at $\sim 2 \times 10^7 \, h^{-1} \, M_\odot$ could be related to the cut, $M_{max} = 3.2 \times 10^7 \, h^{-1} \, M_\odot$, applied as part of the cleaning procedure (Sect. 2.3.3), but, in any case, this is very close to the resolution limit which is also the mass scale at which the spurious haloes begin to dominate the mass function.

The main conclusion to be drawn from Fig. 2.7 is that above the resolution limit, the modified ellipsoidal collapse model reproduces the WDM mass function remarkably accurately, over nearly 6 orders of magnitude in mass.

2.4.3 Halo Density and Mass Profiles

Spherically-averaged radial density profiles provide the simplest and most direct descriptor of halo structure. We calculate profiles in radial shells equally-spaced in $\log(r/r_{200})$. As we discussed in Sect. 2.3.2, haloes of mass above $10^8 \, h^{-1} \, M_\odot$ can be bijectively matched in COCO-WARM and COCO-COLD. To compare density profiles in the two models, we stack the individual profiles of matched and dynamically relaxed haloes in narrow bins of halo mass of width $\Delta \log(M_{200}) = 0.3$. To determine whether or not a halo is relaxed, we make use of the criteria for dynamical equilibrium set out by Neto et al. (2007): (1) the displacement of the centre of mass from the potential centre should be less than $0.07 r_{vir}$ and (2) less than 10% of the mass within r_{vir} should be in the form of substructure.

The stacked differential density profiles are shown in Fig. 2.8 for a variety of mass bins, with the ratio of the densities shown in the bottom panels. For masses sufficiently larger than $\sim 2 \times 10^9 \, h^{-1} \, M_\odot$, we expect negligible differences in the properties of CDM and WDM haloes: this is apparent in mass bins with $M_{200} > 10^{11} \, h^{-1} \, M_\odot$. Systematic differences in the density profiles begin to appear at around $M_{200} \sim 5 \times 10^{10} \, h^{-1} \, M_\odot$: the WDM haloes have slightly but systematically lower central densities than their CDM counterparts. This halo mass is two orders of magnitude higher than the half-mode mass, and an order of magnitude higher than the scale at which the mass functions begin to differ (Fig. 2.7). The difference in central density grows as the mass decreases and reaches $\sim 30\%$ at the smallest mass bin shown, $M_{200} \sim 1.4 \times 10^9 \, h^{-1} \, M_\odot$. We discuss the physical reason for this in the next section.

It is now well established that the density profiles of dark matter haloes in general are well described by the NFW profile (Navarro et al. 1996, 1997):

$$\frac{\rho(r)}{\rho_c} = \frac{\delta_c}{(r/r_s)(1 + r/r_s)^2} , \qquad (2.4.6)$$

where δ_c is a characteristic overdensity and r_s is a scale radius. These two parameters are strongly correlated and depend only on halo mass (Navarro et al. 1997). The NFW form is a nearly universal profile in the sense that it approximately fits the profiles of relaxed haloes of any mass formed by gravitational instability from all the initial

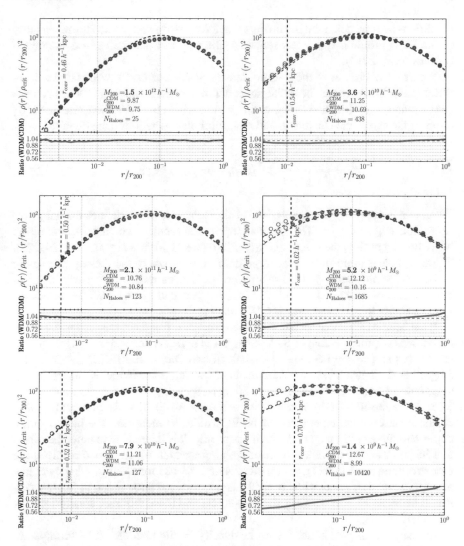

Fig. 2.8 Stacked spherically-averaged density profiles in COCO- WARM (red) and COCO- COLD (blue). For each mass bin we compare the profiles of only relaxed, matched haloes in the two simulations; the number in each bin is indicated in each subpanel. The vertical dashed line represents the convergence radius, $r_{\rm conv}$, and filled symbols indicate the range of the profile above this limit, whereas open symbols denote the radial range below it. The dashed red and blue lines are NFW fits to the WDM and CDM profiles respectively. Note that the density profiles have been scaled by $(r/r_{200})^2$ so as to reduce the dynamic range on the vertical axis. The bottom panels show the ratio of the WDM and CDM densities in each bin. Figure reproduced from Bose et al. (2016)

conditions and cosmological parameters that have been tested so far. The universality of the NFW profile is intimately related to the way in which haloes are assembled (Ludlow et al. 2013).

We fit NFW profiles to the stacked density profiles of COCO- WARM and COCO-COLD in Fig. 2.8, between the radial range defined by the Power et al. (2003) convergence radius, r_{conv} (defined as the radius within which the relaxation time is of the order of the age of the Universe), and r_{200}, minimising the following quantity:

$$\sigma_{\text{fit}}^2 = \frac{1}{N_{\text{bins}} - 1} \sum_{i=1}^{N_{\text{bins}}} [\ln \rho_i - \ln \rho_{\text{NFW}} (\delta_c; r_s)]^2 . \tag{2.4.7}$$

We obtain the best-fitting values of the scale radius, r_s, which defines the halo concentration, $c_{200} = r_{200}/r_s$. This parameter provides a unique characterisation of the NFW density profile; the values of c_{200} for the stacked profiles are quoted in Fig. 2.8. There is a clear trend in that for large halo masses, where the density profiles in COCO- WARM and COCO- COLD are similar, the concentrations are nearly identical but, for masses below $\sim 5 \times 10^{10}$ h$^{-1} M_{\odot}$, the concentrations of WDM haloes are systematically lower than those of CDM haloes.

It has been claimed that for haloes resolved near the free streaming scale, the shape of the density profile is longer NFW-like, but is in fact described by a single power law with a slope ≈ -1.5 (e.g. Anderhalden and Diemand 2013, Ishiyama 2014, Angulo et al. 2017). These measurements have been made at the regime of Earth-mass haloes ($M_{200} \sim 10^{-6} M_{\odot}$) at $z \sim 30$, corresponding to the free streaming scale for a neutralino-like CDM particle with rest mass ≈ 100 GeV. Angulo et al. (2017) attribute the eventual emergence of NFW profiles in more massive haloes to the cumulative impact of mergers on the inner profile. It is interesting to note, therefore, that in the case of the 3.3 keV thermal relic, for which we have also resolved the free streaming scale in this chapter, there is no evidence for steep, single power law profiles at any mass scale, with the NFW shape preserved throughout. In future work, it will be interesting to consider the reasons behind these different conclusions for CDM and WDM.

In many cases, even better fits to the density profile than the NFW formula are provided by a function first used by Einasto (1965) to describe star counts in the Milky Way. This formula, which has an additional free parameter, was dubbed the "Einasto profile" by Navarro et al. (2004), who showed that it provides a very good fit to CDM haloes:

$$\ln \left(\frac{\rho}{\rho_{-2}} \right) = -\frac{2}{\alpha} \left[\left(\frac{r}{r_{-2}} \right)^{\alpha} - 1 \right] , \tag{2.4.8}$$

where ρ_{-2} is the density at $r = r_{-2}$, the radius at which the logarithmic slope of the profile is -2 (or where $r^2 \rho$ has its maximum). The parameter r_{-2} in the Einasto profile is analogous to the scale radius, r_s, of the NFW profile. This allows an equivalent definition of halo concentration, $c_{200} = r_{200}/r_{-2}$. The parameter α (not to be confused with the one in Eq. 2.3.3) is a shape parameter that controls the curvature of the profile in the inner regions. A value of $\alpha \simeq 0.17$ results in a good match to CDM haloes over a wide range of masses (Navarro et al. 2004; Gao et al. 2008).

This is demonstrated in Fig. 2.9, which is similar to Fig. 2.8, but with Einasto profiles fitted instead of NFW profiles. It is apparent that the shape parameter, α, allows a better fit to the halo density profiles in both COCO- WARM and COCO- COLD, especially in the inner parts. It is also interesting to note that the concentrations inferred from the Einasto profile fits tend to be slightly lower than those inferred from the NFW profile fits especially at higher masses.

In Fig. 2.10 we compare the ratio of M_{200} values for individually matched haloes in COCO- WARM and COCO- COLD at the present day. We consider only haloes with $M_{200} > 10^8 \, h^{-1} \, M_\odot$ for which we have almost complete matching ($\sim$97%) between the two simulations, and plot the ratio, $M_{200}^{WDM}/M_{200}^{CDM}$ as a function of M_{200}^{CDM}. The solid red line shows the median ratio, whereas the dashed red lines represent the 16 and 84th percentiles. The masses are very similar for objects $>5 \times 10^{10} \, h^{-1} \, M_\odot$, where the ratios agree to within 1%. For masses lower than this, WDM haloes are systematically less massive than their CDM counterparts, with the deficit in WDM halo mass reaching $\sim$30% at $M_{200}^{CDM} = 10^9 \, h^{-1} \, M_\odot$. Haloes of these masses in WDM form later than their CDM counterparts and thus have less time to grow.

In Fig. 2.11 we show the cumulative radial distribution of mass in haloes in COCO- WARM (red lines) and COCO- COLD (blue squares). The ratios are shown in the lower panels. From Fig. 2.10, we expect the cumulative profiles to be very similar at $r/r_{200} = 1$ except in the lowest mass bin, where WDM haloes are slightly ($\sim$10%) less massive than their CDM matches. The same trend seen in the density profiles is apparent here: for $M_{200} < 5 \times 10^{10} \, h^{-1} \, M_\odot$, the profiles are less concentrated in the central regions in COCO- WARM than in COCO- COLD. The reason for this difference is discussed in the next section.

2.4.4 The Concentration-Mass Relation

As mentioned in the previous section, the density profile of a dark matter halo is characterised by its concentration. As a result of their hierarchical formation process, the inner parts of haloes in CDM and WDM are essentially in place even before the bulk of the halo mass is assembled (Wang et al. 2011). The concentration reflects the mean density of the Universe at the epoch when these inner regions are in place and the earlier a halo forms, the higher its concentration is Navarro et al. (1997).

In Sect. 2.4.3, we found that the Einasto profile provides a slightly better fit to the density profiles of WDM and CDM haloes than does the conventional NFW profile. Furthermore, Einasto fits are less sensitive to the radial fitting range (Gao et al. 2008 but see also Ludlow et al. 2013). For these reasons, we proceed to derive the concentration-mass relation in our simulations using fits of the Einasto profile to the density profiles of *individual* haloes (not the stacks). Again, fitting is performed between the convergence radius, r_{conv}, and r_{200}, while minimising the *rms* of the fit:

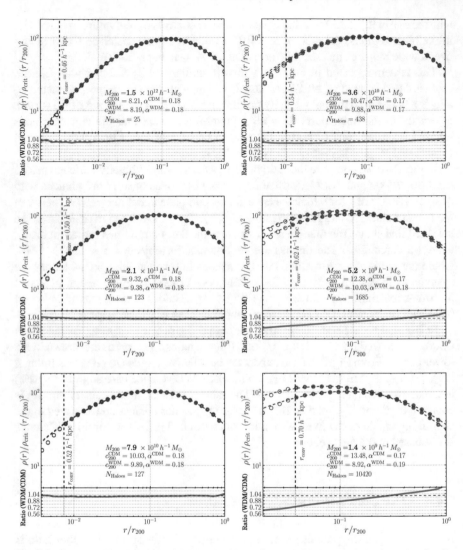

Fig. 2.9 Same as Fig. 2.8, but with Einasto fits to the COCO-WARM and COCO-COLD density profiles. Figure reproduced from Bose et al. (2016)

$$\sigma_{\text{fit}}^2 = \frac{1}{N_{\text{bins}} - 1} \sum_{i=1}^{N_{\text{bins}}} \left[\ln \rho_i - \ln \rho_{\text{Ein}} \left(\rho_{-2}; r_{-2}; \alpha \right) \right]^2 . \tag{2.4.9}$$

To obtain the halo $M_{200} - c_{200}$ relation we first split the haloes into bins equally spaced in logarithmic mass. We then fit an Einasto profile to each halo individually, removing all unrelaxed haloes according to the Neto et al. (2007) criteria. We then

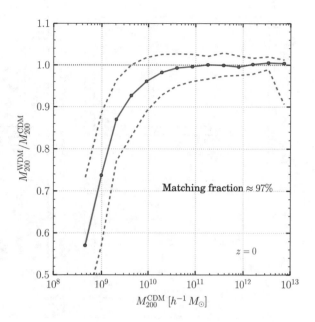

Fig. 2.10 Ratio of halo mass (M_{200}) for all (relaxed and unrelaxed) matched haloes above $M_{200} > 10^8 \, h^{-1} \, M_\odot$ in COCO- WARM and COCO- COLD, as function of M_{200}^{CDM}. The solid red line shows the median relation in bins of M_{200}^{CDM}, whereas the dashed red lines indicate the 16 and 84th percentiles. Figure reproduced from Bose et al. (2016)

find the median halo concentration in each mass bin and estimate its uncertainty using bootstrap resampling.

Figure 2.12 shows the (median) concentration-mass relations for COCO- COLD (dotted lines and shaded regions) and COCO- WARM (points with error bars) at redshifts $z = 0, 0.5, 1, 2, 3$ and 4 (different colours as indicated in the legend). These relations display the same qualitative behaviour seen in the density profiles in Fig. 2.11. For haloes with mass $M_{200} > 10^{11} \, h^{-1} \, M_\odot$, the concentrations of CDM and WDM haloes agree well over all redshifts. For masses below this value, WDM haloes are less concentrated than their CDM counterparts at all redshifts. This is a direct result of the later formation epoch of haloes of a given mass in WDM, and reflects the fact that the mass within r_{-2} in WDM haloes is assembled when the background density of the Universe is lower than in the CDM case.

Whereas the CDM halo concentrations continue to increase as power laws towards lower masses, reflecting hierarchical growth, the WDM halo concentrations turn over at $M_{200} < 5 \times 10^{10} \, h^{-1} \, M_\odot$ and eventually begin to decrease (see also Schneider et al. 2012, Macciò et al. 2013). This echoes the finding in Fig. 2.11 that the mass in the central regions of WDM haloes begins to fall below that in the CDM case roughly below this mass. This mass is an order of magnitude larger than the mass scale at which the mass functions begin to differ ($\sim 2 \times 10^9 \, h^{-1} \, M_\odot$, see Figs. 2.4, 2.7). This result is not entirely surprising: the concentration is sensitive to the inner parts of the profile and it is this inner mass (which we can roughly identify with the matter contained within r_{-2}) which is assembled later in WDM than in CDM, while most of the mass actually lies in the outer parts of the halo.

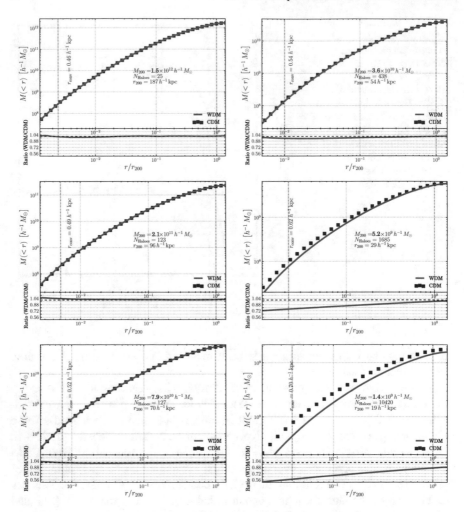

Fig. 2.11 Stacked cumulative mass profiles of relaxed, matched haloes in different mass bins for WDM (solid red lines) and CDM (blue squares). The lower panels show the ratio of the WDM mass to the CDM mass as a function of radius from the centre of the halo (in units of r_{200}). For haloes with $M_{200} > 10^{11}$ $h^{-1} M_\odot$, the mass profiles are nearly identical, but below $M_{200} \leq 5 \times 10^{10}$ $h^{-1} M_\odot$ they differ noticeably. Figure reproduced from Bose et al. (2016)

The lower panel of Fig. 2.12 shows the ratio of the concentrations in COCO-WARM and COCO-COLD, $c_{200}^{WDM}/c_{200}^{CDM}$. There are two interesting features of note: firstly, for all redshifts, the downturn in the WDM halo concentrations occurs at roughly the same halo mass, $M_{200} \sim 5 \times 10^{10}$ $h^{-1} M_\odot$; and secondly, at fixed mass, the ratio decreases with decreasing redshift. The fact that the mass at which WDM halo concentrations begin to peel-off from the CDM relation is almost independent of redshift reflects the narrow redshift range in which the inner parts of WDM haloes form.

Fig. 2.12 The median concentration-mass relation and its redshift evolution for haloes in COCO- COLD and COCO- WARM. The colour dotted lines show the median relation over redshift for CDM haloes, as indicated in the legend. The shaded regions represent the errors in the median, as estimated by bootstrap resampling. The points with the error bars show the corresponding redshift relation in WDM. Only relaxed haloes are included. The thin colour lines show the results of the fitting formula introduced in Eq. 2.4.10. Figure reproduced from Bose et al. (2016)

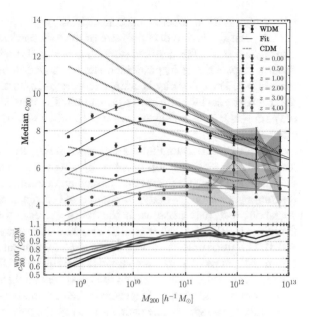

In COCO- WARM we also find that the evolution of the mass-concentration relation over redshift can be approximated using a simple functional form motivated by Eq. 2.4.1 (see Schneider et al. 2012), with an extra redshift-dependent component:

$$\frac{c_{200}^{\mathrm{WDM}}}{c_{200}^{\mathrm{CDM}}} = \left(1 + \gamma_1 \frac{M_{\mathrm{hm}}}{M_{200}}\right)^{-\gamma_2} \times (1+z)^{\beta(z)} . \tag{2.4.10}$$

Here, M_{hm} is the half-mode mass, z is the redshift of interest, $\gamma_1 = 60$, $\gamma_2 = 0.17$ and $\beta(z) = 0.026z - 0.04$. The predictions of our model are shown in the upper panel of Fig. 2.12 using the thin colour lines. While the model does not fully capture the relatively flat relationship at $z = 3$ and 4 in COCO- WARM, it generally reproduces the trends in the simulation and provides a good fit up to $z = 2$, over nearly 5 orders of magnitude in halo mass.

2.4.5 The Shapes and Spins of Haloes

In this section we examine the shapes and spins of WDM haloes. The shapes are most commonly quantified by the *triaxiality*, defined through the halo inertia tensor:

$$I_{ij} = m_p \sum_{n=1}^{N_{200}} x_{n,i} x_{n,j} , \tag{2.4.11}$$

where N_{200} is the number of particles within r_{200}, m_p is the mass of the simulation particle, and $x_{n,i}$ is the ith coordinate of the n-th particle relative to the halo centre. The eigenvalues of the inertia tensor define the axial lengths of an equivalent uniform density ellipsoid, $a \geq b \geq c$, which can be related to those of the halo itself (Bett et al. 2007). The sphericity is defined as c/a (as in Sect. 2.3.3): the higher its value, the less aspherical the ellipsoid's projection. The triaxiality is defined as $T = (a^2 - b^2)/(a^2 - c^2)$: large values correspond to prolate ellipsoids, small values to oblate ellipsoids.

The results for our simulations are shown in Fig. 2.13, where blue represents CDM and red WDM, with the top panel comparing the median triaxiality, and the lower the median sphericity. Errors on the median quantities were obtained by bootstrap resampling. Previous N-body simulations of CDM haloes have shown that triaxiality correlates with halo mass, with triaxiality decreasing with decreasing halo mass (Frenk et al. 1988; Allgood et al. 2006; Muñoz-Cuartas et al. 2011; Macciò et al. 2013). This trend reflects, in part, the younger dynamical age of more massive haloes. Figure 2.13 shows that the same trend is present for WDM haloes but below $M_{200} \sim 10^{10}$ h^{-1} $M_\odot$, WDM haloes are slightly less triaxial than their CDM counterparts.

A more significant trend is revealed when comparing the spin of haloes in the two simulations. The spin is best characterised by the parameter, λ, defined as:

$$\lambda = \frac{J\sqrt{|E|}}{GM^{5/2}} \qquad (2.4.12)$$

Fig. 2.13 Median halo triaxiality (top panel) and halo sphericity (lower panel) in COCO- WARM (red points) and COCO- COLD (blue lines). The errors on the median were obtained by bootstrapping 100 different samples in each case, and is represented by the red error bars for WDM and the blue shaded region for CDM. Only particles within r_{200} were used to compute these properties from the inertia tensor. Figure reproduced from Bose et al. (2016)

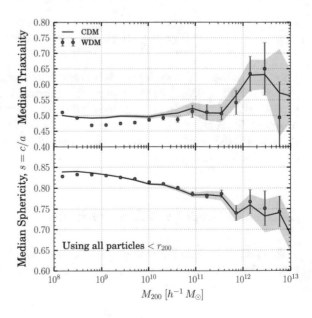

Peebles (1969), where J is the magnitude of the angular momentum of the halo, E is its total energy and M is the mass (which we take to be M_{200}). Haloes acquire a net angular momentum through tidal torques during growth in the linear regime which can be subsequently modified and rearranged by mergers (Peebles 1969; Doroshkevich 1970; White 1984). Since the merger histories are different for CDM and WDM haloes, we might expect some differences in their final angular momentum configurations. In particular, given that tidal forces associated with mergers tend to redistribute angular momentum from the central parts of haloes to the rest of the halo, the smaller frequency of mergers in WDM might facilitate the formation of extended spinning galactic disks (Frenk et al. 1988; Navarro and Benz 1991; Navarro and White 1994).[4]

The spin parameters in our two simulations are compared in the top panel of Fig. 2.14. Previous cosmological CDM simulations showed a very weak correlation between spin and halo mass, with a median value of $\lambda \approx 0.033$, across a wide range of halo masses (Davis et al. 1985; Barnes and Efstathiou 1987; Warren et al. 1992; Steinmetz and Bartelmann 1995; Cole and Lacey 1996; Mo et al. 1998; Bett et al. 2007). Our COCO- COLD simulation reproduces this trend and extends it to lower masses, $M_{200} = 10^8 \, h^{-1} M_\odot$.

For $M_{200} > 5 \times 10^{10} \, h^{-1} M_\odot$, the λ values for WDM haloes are almost identical to those of their CDM counterparts. However, for smaller halo masses λ decreases systematically with decreasing mass and is lower than the CDM value by almost 30% at $M_{200} \sim 10^8 \, h^{-1} M_\odot$. This is consistent with the results of Bullock et al. (2002), who found that three out of four haloes below the WDM cutoff in their simulation had lower values of λ than the equivalent CDM matches. Note that in the top panel of Fig. 2.14 we include all haloes, not necessarily matches, which explains why in some of the largest mass bins, the median spins are not exactly the same in WDM and CDM. In addition, we only include haloes with more than 1000 particles within r_{200} since particle shot noise dominates the estimates of angular momentum for low particle numbers (Frenk et al. 1988; Bett et al. 2007, although we use a more conservative lower limit than the latter's choice of 300 particles).

To investigate why the spins of dwarf galaxy haloes are lower in WDM than in CDM we consider the relative contributions of energy, angular momentum and M_{200} to λ, illustrated in the bottom panel of Fig. 2.14, this time for bijectively matched haloes. The ratio of the median spin parameters is shown by the black squares and the ratio of the geometric means of the quantities that enter into Eq. 2.4.12 are shown by the other colour lines (magenta for J_{CDM}/J_{WDM}, cyan for $|E_{CDM}/E_{WDM}|^{1/2}$ and yellow for $\left(M_{200,CDM}/M_{200,WDM}\right)^{5/2}$). The combination of these ratios in Eq. 2.4.12 should reproduce the ratio of spin parameters, and this is shown in the thick green line. Part of the reason for lower WDM spins below $\sim 10^{10} \, h^{-1} M_\odot$ is their slightly lower total energy which results from their lower concentration. The dominant fac-

---

[4]We note that the inability of many early simulations to form extended disks in the CDM model – the so-called "angular momentum" problem – is readily solved when appropriate prescriptions for supernovae feedback are included in the simulations (see e.g. Okamoto et al. 2005, Scannapieco et al. 2011).

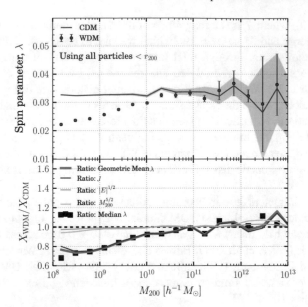

Fig. 2.14 Top panel: the median halo spin-mass relation at $z = 0$ for COCO- WARM (red points) and COCO- COLD (blue line). Errors on the median for the WDM (shown by error bars) and for CDM (shown as the shaded region) haloes were calculated by bootstrap resampling. Bottom panel: the relative contributions of energy, angular momentum and halo mass to the spin of the halo. The black squares show the ratio (CDM to WDM) of the median spin parameters (from the top panel). The magenta, cyan and yellow lines show the ratios of the geometric means of the angular momentum, energy and M_{200} respectively, which when multiplied together appropriately yield the thick green line, which show the ratio of the geometric means of λ_{CDM} and λ_{WDM}. As expected, the squares trace out the ratio of the geometric means. Note that ratios of all quantities are taken between the bijectively matched COCO- WARM and COCO- COLD haloes. Figure reproduced from Bose et al. (2016)

tor, however, is their lower angular momentum relative to CDM haloes, $\sim 25\%$ at $10^8 \ h^{-1} \ M_\odot$. The cause of this could be related to the differing merger histories in WDM and CDM and the likely more quiescent mass accretion of WDM haloes which can result in smaller spins (Bullock et al. 2002; Vitvitska et al. 2002; Hetznecker and Burkert 2006).

2.5 Summary and Discussions

We have presented results from the *Copernicus Complexio* project, a set of cosmological "zoom" simulations in which the dark matter is assumed to be either CDM (COCO- COLD) or a thermal 3.3 keV WDM particle (COCO- WARM). The combination of mass resolution and volume of our simulations provides a rich statistical sample of haloes over seven decades in mass. This WDM model is particularly interesting because it corresponds to the "warmest" particle allowed by current Lyman-α constraints (Viel et al. 2013) and has a linear power spectrum cutoff similar to that for the "coldest" 7 keV sterile neutrino, evidence for which has recently been claimed to be found in galaxies and clusters (Bulbul et al. 2014b; Boyarsky et al. 2014). This cutoff – manifest in haloes of $M_{200} \lesssim 2 \times 10^9 \, h^{-1} \, M_\odot$ for our assumed particle mass – is reflected both in the population statistics and the structure of individual haloes.

The formation of structure begins significantly later in COCO- WARM than in COCO- COLD. Across all redshifts, differences in the halo mass function between COCO- WARM and COCO- COLD begin to appear at a mass roughly one order magnitude larger than the nominal half-mode mass. Below $\sim 2 \times 10^9 \, h^{-1} \, M_\odot$, the WDM mass function declines rapidly but there are still some small haloes present at surprisingly large redshifts: at $z = 10$, for example, there are almost 5 times as many haloes with $M_{200} \sim 10^8 \, h^{-1} \, M_\odot$ in COCO- COLD than in COCO- WARM. We find that the $z = 0$ halo mass functions in both COCO- WARM and COCO- COLD are well described by previous analytic fits to the CDM halo mass function (e.g. Sheth and Tormen 1999) down to our resolution limit, $M_{200} \sim 3 \times 10^7 \, h^{-1} \, M_\odot$, provided that the window function used to compute the mass variance, $\sigma^2(M)$, in the WDM case is calculated using a sharp k-space filter, as described by Benson et al. (2013).

Just as for COCO- COLD, the spherically averaged density profiles of haloes in COCO- WARM, down to dwarf galaxy scales, are well described by NFW or Einasto profiles. The concentration-mass relation, $M_{200} - c_{200}$ (where we have defined concentration using the Einasto profile), in COCO- WARM begins to peel off from the corresponding relation in COCO- COLD at a mass of $\sim 5 \times 10^{10} \, h^{-1} \, M_\odot$, reflecting the later formation epoch of haloes of a given mass in WDM compared to CDM. This mass is larger than the scale below which the WDM mass function is suppressed because halo concentration is determined by the epoch when the *inner* regions of a halo form. The mass at which the concentration begins to differ in the two simulations is almost independent of redshift out to $z \simeq 4$. At the present day, the typical concentration of a halo of mass $10^9 \, h^{-1} \, M_\odot$ in COCO- WARM is $c_{200} \simeq 8$ compared to $c_{200} \simeq 12.7$ in COCO- COLD. The trends and evolution of the concentration-mass relation can be approximated by the fitting formula provided in Eq. 2.4.10.

The generally triaxial shapes of haloes in COCO- WARM and COCO- COLD are very similar. However, we find that, for masses below $\sim 5 \times 10^{10} \, h^{-1} \, M_\odot$, WDM haloes have slightly lower values of the spin parameter, λ, (up to 30%) than their CDM counterparts.

In principle, gravitational lensing is one of the most promising techniques for distinguishing between WDM and CDM, as it directly probes the halo mass function (see for example Vegetti and Koopmans 2009). In the parent volumes of the COCO simulations, the non-linear power spectrum, $P(k)$, for COLOR- WARM is suppressed by $\sim 3\%$ relative to COLOR- COLD on scales $k \lesssim 5 \, h \, \mathrm{Mpc}^{-1}$ (consistent with the simulations of Viel et al. 2012, which bracket the 3.3 keV model). While the weak lensing signal on these scales should be measurable by surveys such as DESI and EUCLID, this difference is smaller than the differences introduced by baryon effects on the dark matter-only $P(k)$, which is of the order of $5 - 10\%$, as seen in hydrodynamic simulations (van Daalen et al. 2014; van Daalen and Schaye 2015; Hellwing et al.

2016b). It is therefore necessary to use hydrodynamic simulations to check for any residual signal of the nature of the dark matter species, both in the power spectrum and in other observable properties of the galaxy population.

2.6 New Developments Since Submission of This Thesis in April 2017

In Sect. 2.4.4, we presented the concentration-mass relation and its evolution with redshift for CDM and WDM haloes in the COCO simulations (Fig. 2.12). Equation 2.4.10 provides simple fitting functions that describe these relations. Of course, the coefficients in these equations have been obtained through fits to COCO and are specific to a 3.3 keV thermal relic (or models that exhibit a cutoff at a similar scale).

A more sophisticated model for concentrations in WDM was developed in Ludlow et al. (2016). This work builds on the findings by Ludlow et al. (2013), who realised an intimate connection between the shape of the density profile and assembly history of dark matter haloes. In particular, when expressed in terms of the "collapsed mass history" i.e., the total mass of all its collapsed progenitors, the assembly history of WDM haloes – either above or below the cutoff – exhibit a universal form. Ludlow et al. (2016) relate the mean enclosed density within the scale radius, $\langle \rho_{-2} \rangle$, to the critical density of the Universe at the collapse redshift of the halo, z_{-2}, using the scaling relation:

$$\frac{\langle \rho_{-2} \rangle}{\rho_0} = C \times \frac{\rho_c(z_{-2})}{\rho_0} \,, \qquad (2.6.1)$$

where ρ_0 is the critical density during the epoch at which the halo is identified and $C = 650$. The collapse redshift is defined as the redshift at which the mass within the scale radius, M_{-2}, was first contained in progenitors more massive than 2% of the final halo mass. For an Einasto profile, $\langle \rho_{-2} \rangle$ can then be used to determine the concentration of the halo by solving the following equation:

$$\langle \rho(r) \rangle = \frac{200}{x^3} \frac{\Gamma\,(3/\alpha; 2/\alpha(xc)^\alpha)}{\Gamma\,(3/\alpha; 2/\alpha c^\alpha)} \rho_c \,, \qquad (2.6.2)$$

where α is again the shape parameter of the Einasto profile, $x = r/r_{200}$ and $\Gamma(a; y)$ is the incomplete gamma function. Predictions of this model with the mass-concentration relations presented in Fig. 2.12 are shown in Fig. 2.15. The analytic model described above provides a remarkably good match to the relations obtained from the simulation over a wide range of halo mass and redshift.

The great benefit of such a model is its predictability: by simply specifying the linear power spectrum, it is possible to use the machinery linking the collapsed mass history to mass profiles to predict the concentration-mass-redshift relation. The model we present in Ludlow et al. (2016) was shown to provide a good match

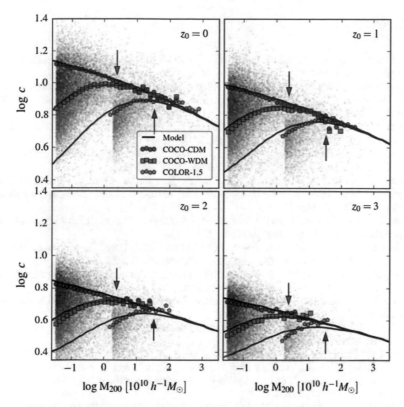

Fig. 2.15 Mass-concentration-redshift relation for COCO- COLD (blue), COCO- WARM and the COLOR WDM simulation with a relic mass of 1.5 keV. Fits to individual haloes are show as coloured dots; heavy points correspond to the best-fit relation derived from the *stacked* mass profiles of haloes in equally-spaced logarithmic mass bins of width 0.1 dex. Only bins containing at least 25 haloes are shown. The arrows indicate a mass scale corresponding to one hundred times the half-mode mass. The solid lines correspond to the predictions of the analytic model of Ludlow et al. 2016. Figure reproduced from Ludlow et al. (2016)

to the results of simulations spanning a wide range of cosmological parameters and linear theory cutoffs. Note, however, the predictability of this model is limited to the case where dark matter is collisionless. Particle physics scenarios where the dark matter undergoes elastic/inelastic scattering (self-interacting dark matter, e.g. Spergel and Steinhardt 2000) result in density profiles that are shaped by dynamical processes beyond just the accretion history. Models for predicting the concentration-mass relation in these theories require a prescription that additionally predicts the scattering rate of dark matter particles and the subsequent redistribution of mass in haloes.

References

Abazajian KN (2014) Phys Rev Lett 112:161303. https://doi.org/10.1103/PhysRevLett.112. 161303, http://adsabs.harvard.edu/abs/2014PhRvL.112p1303A

Allgood B, Flores RA, Primack JR, Kravtsov AV, Wechsler RH, Faltenbacher A, Bullock JS (2006) MNRAS 367:1781. https://doi.org/10.1111/j.1365-2966.2006.10094.x, http://adsabs.harvard.edu/abs/2006MNRAS.367.1781A

Anderhalden D, Diemand J (2013) JCAP 4:009. https://doi.org/10.1088/1475-7516/2013/04/009, http://adsabs.harvard.edu/abs/2013JCAP...04..009A

Anderson ME, Churazov E, Bregman JN (2015) MNRAS 452:3905. https://doi.org/10.1093/mnras/stv1559, http://adsabs.harvard.edu/abs/2015MNRAS.452.3905A

Angulo RE, Springel V, White SDM, Jenkins A, Baugh CM, Frenk CS (2012) MNRAS 426:2046. https://doi.org/10.1111/j.1365-2966.2012.21830.x, http://adsabs.harvard.edu/abs/2012MNRAS.426.2046A

Angulo RE, Hahn O, Abel T (2013) MNRAS 434:3337. https://doi.org/10.1093/mnras/stt1246, http://adsabs.harvard.edu/abs/2013MNRAS.434.3337A

Angulo RE, Hahn O, Ludlow AD, Bonoli S (2017) MNRAS 471:4687. https://doi.org/10.1093/mnras/stx1658, http://adsabs.harvard.edu/abs/2017MNRAS.471.4687A

Asaka T, Shaposhnikov M (2005) Phys. Lett. B 620:17. https://doi.org/10.1016/j.physletb.2005.06.020, http://adsabs.harvard.edu/abs/2005PhLB..620...17A

Barnes J, Efstathiou G (1987) ApJ 319:575. https://doi.org/10.1086/165480, http://adsabs.harvard.edu/abs/1987ApJ...319..575B

Benson AJ et al (2013) MNRAS 428:1774. https://doi.org/10.1093/mnras/sts159, http://adsabs.harvard.edu/abs/2013MNRAS.428.1774B

Bett P, Eke V, Frenk CS, Jenkins A, Helly J, Navarro J (2007) MNRAS 376:215. https://doi.org/10.1111/j.1365-2966.2007.11432.x, http://adsabs.harvard.edu/abs/2007MNRAS.376..215B

Bode P, Ostriker JP, Turok N (2001) ApJ 556:93. https://doi.org/10.1086/321541, http://adsabs.harvard.edu/abs/2001ApJ...556...93B

Bond JR, Szalay AS (1983) ApJ 274:443. https://doi.org/10.1086/161460, http://adsabs.harvard.edu/abs/1983ApJ...274..443B

Bond JR, Cole S, Efstathiou G, Kaiser N (1991) ApJ 490:493. https://doi.org/10.1086/170520, http://adsabs.harvard.edu/abs/1991ApJ...379..440B

Bose S, Hellwing WA, Frenk CS, Jenkins A, Lovell MR, Helly JC, Li B (2016) MNRAS 455:318. https://doi.org/10.1093/mnras/stv2294, http://adsabs.harvard.edu/abs/2016MNRAS.455..318B

Boyarsky A, Ruchayskiy O, Shaposhnikov M (2009) Annu Rev Nucl Part Sci 59:191. https://doi.org/10.1146/annurev.nucl.010909.083654, http://adsabs.harvard.edu/abs/2009ARNPS..59..191B

Boyarsky A, Ruchayskiy O, Iakubovskyi D, Franse J (2014) Phys Rev Lett 113:251301. https://doi.org/10.1103/PhysRevLett.113.251301, http://adsabs.harvard.edu/abs/2014PhRvL.113y1301B

Boyarsky A, Franse J, Iakubovskyi D, Ruchayskiy O (2015) Phys Rev Lett 115:161301. https://doi.org/10.1103/PhysRevLett.115.161301, http://adsabs.harvard.edu/abs/2015PhRvL.115p1301B

Bryan GL, Norman ML (1998) ApJ 495:80. https://doi.org/10.1086/305262, http://adsabs.harvard.edu/abs/1998ApJ...495...80B

Bulbul E, Markevitch M, Foster AR, Smith RK, Loewenstein M, Randall SW (2014a). arXiv:1409.4143

Bulbul E, Markevitch M, Foster A, Smith RK, Loewenstein M, Randall SW (2014b) ApJ 789:13. https://doi.org/10.1088/0004-637X/789/1/13, http://adsabs.harvard.edu/abs/2014ApJ...789...13B

Bullock JS, Kravtsov aAV, Colín P (2002) ApJL 564:L1. https://doi.org/10.1086/338863, http://adsabs.harvard.edu/abs/2002ApJ...564L...1B

Cole S, Lacey C (1996) MNRAS 281:716. http://adsabs.harvard.edu/abs/1996MNRAS.281..716C

Cole S et al (2005) MNRAS 362:505. https://doi.org/10.1111/j.1365-2966.2005.09318.x, http://adsabs.harvard.edu/abs/2005MNRAS.362..505C

Colín P, Valenzuela O, Avila-Reese V (2008) ApJ 673:203. https://doi.org/10.1086/524030, http://adsabs.harvard.edu/abs/2008ApJ...673..203C

Colín P, Avila-Reese V, González-Samaniego A, Velázquez H (2015) ApJ 803:28. https://doi.org/10.1088/0004-637X/803/1/28, http://adsabs.harvard.edu/abs/2015ApJ...803...28C

Colless M et al (2001) MNRAS 328:1039. https://doi.org/10.1046/j.1365-8711.2001.04902.x, http://adsabs.harvard.edu/abs/2001MNRAS.328.1039C

Conlon JP, Day FV (2014) JCAP 11:33. https://doi.org/10.1088/1475-7516/2014/11/033, http://adsabs.harvard.edu/abs/2014JCAP...11..033C

Crain RA et al (2009) MNRAS 399:1773. https://doi.org/10.1111/j.1365-2966.2009.15402.x, http://adsabs.harvard.edu/abs/2009MNRAS.399.1773C

Davis M, Efstathiou G, Frenk CS, White SDM (1985) ApJ 292:371. https://doi.org/10.1086/163168, http://adsabs.harvard.edu/abs/1985ApJ...292..371D

Dodelson S, Widrow LM (1994) Phys Rev Lett 72:17. https://doi.org/10.1103/PhysRevLett.72.17, http://adsabs.harvard.edu/abs/1994PhRvL..72...17D

Doroshkevich AG (1970) Astrophysics 6:320. https://doi.org/10.1007/BF01001625, http://adsabs.harvard.edu/abs/1970Ap......6..320D

Einasto J (1965) Trudy Astrofizicheskogo Instituta Alma-Ata 5:87. http://adsabs.harvard.edu/abs/1965TrAlm...5...87E

Eisenstein DJ et al (2005) ApJ 633:560.https://doi.org/10.1086/466512, http://adsabs.harvard.edu/abs/2005ApJ...633..560E

Eke VR, Cole S, Frenk CS (1996) MNRAS 282:263. http://adsabs.harvard.edu/abs/1996MNRAS.282..263E

Frenk CS, White SDM (2012) Annalen der Physik 524:507. https://doi.org/10.1002/andp.201200212, http://adsabs.harvard.edu/abs/2012AnP...524..507F

Frenk CS, White SDM, Efstathiou G, Davis M (1985) Nature 317:595. https://doi.org/10.1038/317595a0, http://adsabs.harvard.edu/abs/1985Natur.317..595F

Frenk CS, White SDM, Davis M, Efstathiou G (1988) ApJ 327:507. https://doi.org/10.1086/166213, http://adsabs.harvard.edu/abs/1988ApJ...327..507F

Frenk CS, Evrard AE, White SDM, Summers FJ (1996) ApJ 472:460. https://doi.org/10.1086/178079, http://adsabs.harvard.edu/abs/1996ApJ...472..460F

Gao L, Navarro JF, Cole S, Frenk CS, White SDM, Springel V, Jenkins A, Neto AF (2008) MNRAS 387:536. https://doi.org/10.1111/j.1365-2966.2008.13277.x, http://adsabs.harvard.edu/abs/2008MNRAS.387..536G

Hahn O, Abel T, Kaehler R (2013) MNRAS 434:1171.https://doi.org/10.1093/mnras/stt1061, http://adsabs.harvard.edu/abs/2013MNRAS.434.1171H

Harker G, Cole S, Helly J, Frenk C, Jenkins A (2006) MNRAS 367:1039. https://doi.org/10.1111/j.1365-2966.2006.10022.x, http://adsabs.harvard.edu/abs/2006MNRAS.367.1039H

Hawkins E et al (2003) MNRAS 346:78.https://doi.org/10.1046/j.1365-2966.2003.07063.x, http://adsabs.harvard.edu/abs/2003MNRAS.346...78H

Hellwing WA, Frenk CS, Cautun M, Bose S, Helly J, Jenkins A, Sawala T, Cytowski M (2016a) MNRAS 457:3492. https://doi.org/10.1093/mnras/stw214, http://adsabs.harvard.edu/abs/2016MNRAS.457.3492H

Hellwing WA, Schaller M, Frenk CS, Theuns T, Schaye J, Bower RG, Crain RA (2016b) MNRAS 461:L11. https://doi.org/10.1093/mnrasl/slw081, http://adsabs.harvard.edu/abs/2016MNRAS.461L..11H

Hetznecker H, Burkert A (2006) MNRAS 370:1905. https://doi.org/10.1111/j.1365-2966.2006.10616.x, http://adsabs.harvard.edu/abs/2006MNRAS.370.1905H

Hobbs A, Read JI, Agertz O, Iannuzzi F, Power C (2016) MNRAS 458:468. https://doi.org/10.1093/mnras/stw251, http://adsabs.harvard.edu/abs/2016MNRAS.458..468H

Hooper D, Goodenough L (2011) Physics Letters B 697:412. https://doi.org/10.1016/j.physletb.2011.02.029, http://adsabs.harvard.edu/abs/2011PhLB..697..412H

Ishiyama T (2014) ApJ 788:27. https://doi.org/10.1088/0004-637X/788/1/27, http://adsabs.harvard.edu/abs/2014ApJ...788...27I

Jeltema T, Profumo S (2015) MNRAS 450:2143. https://doi.org/10.1093/mnras/stv768, http://
    adsabs.harvard.edu/abs/2015MNRAS.450.2143J
Jenkins A (2010) MNRAS 403:1859. https://doi.org/10.1111/j.1365-2966.2010.16259.x, http://
    adsabs.harvard.edu/abs/2010MNRAS.403.1859J
Jenkins A (2013) MNRAS 434:2094. https://doi.org/10.1093/mnras/stt1154, http://adsabs.harvard.
    edu/abs/2013MNRAS.434.2094J
Jenkins A, Frenk CS, White SDM, Colberg JM, Cole S, Evrard AE, Couchman HMP, Yoshida
    N (2001) MNRAS 321:372. https://doi.org/10.1046/j.1365-8711.2001.04029.x, http://adsabs.
    harvard.edu/abs/2001MNRAS.321..372J
Katz N, White SDM (1993) ApJ 412:455. https://doi.org/10.1086/172935, http://adsabs.harvard.
    edu/abs/1993ApJ...412..455K
Komatsu E et al (2011) APJS 192:18. https://doi.org/10.1088/0067-0049/192/2/18, http://adsabs.
    harvard.edu/abs/2011ApJS..192...18K
Lacey C, Cole S (1993) MNRAS 262:627. http://adsabs.harvard.edu/abs/1993MNRAS.262..627L
Laine M, Shaposhnikov M (2008) JCAP 6:31. https://doi.org/10.1088/1475-7516/2008/06/031,
    http://adsabs.harvard.edu/abs/2008JCAP...06..031L
Li R, Frenk CS, Cole S, Gao L, Bose S, Hellwing WA (2016) MNRAS 460:363. https://doi.org/10.
    1093/mnras/stw939, http://adsabs.harvard.edu/abs/2016MNRAS.460..363L
Lovell MR et al (2012) MNRAS 420:2318. https://doi.org/10.1111/j.1365-2966.2011.20200.x,
    http://adsabs.harvard.edu/abs/2012MNRAS.420.2318L
Lovell MR, Frenk CS, Eke VR, Jenkins A, Gao L, Theuns T (2014) MNRAS 439:300. https://doi.
    org/10.1093/mnras/stt2431, http://adsabs.harvard.edu/abs/2014MNRAS.439..300L
Lovell MR et al (2016) MNRAS 461:60. https://doi.org/10.1093/mnras/stw1317, http://adsabs.
    harvard.edu/abs/2016MNRAS.461...60L
Ludlow AD et al (2013) MNRAS 432:1103.https://doi.org/10.1093/mnras/stt526, http://adsabs.
    harvard.edu/abs/2013MNRAS.432.1103L
Ludlow AD, Bose S, Angulo RE, Wang L, Hellwing WA, Navarro JF, Cole S, Frenk CS
    (2016) MNRAS 460:1214. https://doi.org/10.1093/mnras/stw1046, http://adsabs.harvard.edu/
    abs/2016MNRAS.460.1214L
Macciò AV, Paduroiu S, Anderhalden D, Schneider A, Moore B (2012) MNRAS
    428:1774. https://doi.org/10.1111/j.1365-2966.2012.21284.x, http://adsabs.harvard.edu/abs/
    2012MNRAS.424.1105M
Macciò AV, Ruchayskiy O, Boyarsky A, Muñoz-Cuartas JC (2013) MNRAS 428:882. https://doi.
    org/10.1093/mnras/sts078, http://adsabs.harvard.edu/abs/2013MNRAS.428..882M
Malyshev D, Neronov A, Eckert D (2014) Phys Rev D 90:103506. https://doi.org/10.1103/
    PhysRevD.90.103506, http://adsabs.harvard.edu/abs/2014PhRvD..90j3506M
Menci N, Fiore F, Lamastra A (2012) MNRAS 421:2384. https://doi.org/10.1111/j.1365-2966.
    2012.20470.x, http://adsabs.harvard.edu/abs/2012MNRAS.421.2384M
Mo HJ, Mao S, White SDM (1998) MNRAS 295:319. https://doi.org/10.1046/j.1365-8711.1998.
    01227.x, http://adsabs.harvard.edu/abs/1998MNRAS.295..319M
Muñoz-Cuartas JC, Macciò AV, Gottlöber S, Dutton AA (2011) MNRAS 411:584. https://doi.org/
    10.1111/j.1365-2966.2010.17704.x, http://adsabs.harvard.edu/abs/2011MNRAS.411..584M
Navarro JF, Benz W (1991) ApJ 380:320. https://doi.org/10.1086/170590, http://adsabs.harvard.
    edu/abs/1991ApJ...380..320N
Navarro JF, White SDM (1994) MNRAS 267:401. http://adsabs.harvard.edu/abs/1994MNRAS.
    267..401N
Navarro JF, Frenk CS, White SDM (1996) ApJ 462:563. https://doi.org/10.1086/177173, http://
    adsabs.harvard.edu/abs/1996ApJ...462..563N
Navarro JF, Frenk CS, White SDM (1997) ApJ 490:493. http://adsabs.harvard.edu/abs/1997ApJ...
    490..493N
Navarro JF et al (2004) MNRAS 349:1039. https://doi.org/10.1111/j.1365-2966.2004.07586.x,
    http://adsabs.harvard.edu/abs/2004MNRAS.349.1039N

Neto AF et al (2007) MNRAS 381:1450. https://doi.org/10.1111/j.1365-2966.2007.12381.x, http://adsabs.harvard.edu/abs/2007MNRAS.381.1450N

Okamoto T, Eke VR, Frenk CS, Jenkins A (2005) MNRAS 363:1299. https://doi.org/10.1111/j.1365-2966.2005.09525.x, http://adsabs.harvard.edu/abs/2005MNRAS.363.1299O

Peebles PJE (1969) ApJ 155:393. https://doi.org/10.1086/149876, http://adsabs.harvard.edu/abs/1969ApJ...155..393P

Planck Collaboration et al (2014) A&A 571:A1. https://doi.org/10.1051/0004-6361/201321529, http://adsabs.harvard.edu/abs/2014A%26A...571A...1P

Power C, Navarro JF, Jenkins A, Frenk CS, White SDM, Springel V, Stadel J, Quinn T (2003) MNRAS 338:14. https://doi.org/10.1046/j.1365-8711.2003.05925.x, http://adsabs.harvard.edu/abs/2003MNRAS.338...14P

Press WH, Schechter P (1974) ApJ 187:425. https://doi.org/10.1086/152650, http://adsabs.harvard.edu/abs/1974ApJ...187..425P

Reed DS, Bower R, Frenk CS, Jenkins A, Theuns T (2007) MNRAS 374:2. https://doi.org/10.1111/j.1365-2966.2006.11204.x, http://adsabs.harvard.edu/abs/2007MNRAS.374....2R

Riemer-Sørensen S (2016) A&A 590:A71. https://doi.org/10.1051/0004-6361/201527278, http://adsabs.harvard.edu/abs/2016A%26A...590A..71R

Sawala T, Frenk CS, Crain RA, Jenkins A, Schaye J, Theuns T, Zavala J (2013) MNRAS 431:1366. https://doi.org/10.1093/mnras/stt259, http://adsabs.harvard.edu/abs/2013MNRAS.431.1366S

Sawala T et al (2015) MNRAS 431:1366. https://doi.org/10.1093/mnras/stu2753, http://adsabs.harvard.edu/abs/2015MNRAS.448.2941S

Sawala T, et al (2016) MNRAS 431:1366. https://doi.org/10.1093/mnras/stv2597, http://adsabs.harvard.edu/abs/2016MNRAS.456...85S

Scannapieco C, White SDM, Springel V, Tissera PB (2011) MNRAS 417:154. https://doi.org/10.1111/j.1365-2966.2011.19027.x, http://adsabs.harvard.edu/abs/2011MNRAS.417..154S

Schaller M et al (2015) MNRAS 431:1366. https://doi.org/10.1093/mnras/stv1067, http://adsabs.harvard.edu/abs/2015MNRAS.451.1247S

Schneider A, Smith RE, Macciò A. V., Moore B (2012) MNRAS 424:684. https://doi.org/10.1111/j.1365-2966.2012.21252.x, http://adsabs.harvard.edu/abs/2012MNRAS.424..684S

Schneider A, Smith RE, Reed D (2013) MNRAS 433:1573. https://doi.org/10.1093/mnras/stt829, http://adsabs.harvard.edu/abs/2013MNRAS.433.1573S

Shao S, Gao L, Theuns T, Frenk CS (2013) MNRAS 430:2346. https://doi.org/10.1093/mnras/stt053, http://adsabs.harvard.edu/abs/2013MNRAS.430.2346S

Sheth RK, Tormen G (1999) MNRAS 308:119. https://doi.org/10.1046/j.1365-8711.1999.02692.x, http://adsabs.harvard.edu/abs/1999MNRAS.308..119S

Shi X, Fuller GM (1999) Physical Review Letters 82:2832. https://doi.org/10.1103/PhysRevLett.82.2832, http://adsabs.harvard.edu/abs/1999PhRvL..82.2832S

Spergel DN, Steinhardt PJ (2000) Phys Rev Lett 84:3760. https://doi.org/10.1103/PhysRevLett.84.3760, http://adsabs.harvard.edu/abs/2000PhRvL..84.3760S

Springel V (2005) MNRAS 364:1105. https://doi.org/10.1111/j.1365-2966.2005.09655.x, http://adsabs.harvard.edu/abs/2005MNRAS.364.1105S

Springel V, Yoshida N, White SDM (2001a) Nature 6:79. https://doi.org/10.1016/S1384-1076(01)00042-2, http://adsabs.harvard.edu/abs/2001NewA....6...79S

Springel V, White SDM, Tormen G, Kauffmann G (2001b) MNRAS 328:726. https://doi.org/10.1046/j.1365-8711.2001.04912.x, http://adsabs.harvard.edu/abs/2001MNRAS.328..726S

Steinmetz M, Bartelmann M (1995) MNRAS 272:570. http://adsabs.harvard.edu/abs/1995MNRAS.272..570S

Tegmark M et al (2004) Phys Rev D 69:103501. https://doi.org/10.1103/PhysRevD.69.103501, http://adsabs.harvard.edu/abs/2004PhRvD..69j3501T

Tinker J, Kravtsov AV, Klypin A, Abazajian K, Warren M, Yepes G, Gottlöber S, Holz DE (2008) ApJ 688:709. https://doi.org/10.1086/591439, http://adsabs.harvard.edu/abs/2008ApJ...688..709T

Urban O, Werner N, Allen SW, Simionescu A, Kaastra JS, Strigari LE (2015) MNRAS 451:2447. https://doi.org/10.1093/mnras/stv1142, http://adsabs.harvard.edu/abs/2015MNRAS.451.2447U

van Daalen MP, Schaye J (2015) MNRAS 452:2247. https://doi.org/10.1093/mnras/stv1456, http://adsabs.harvard.edu/abs/2015MNRAS.452.2247V

van Daalen MP, Schaye J, McCarthy IG, Booth CM, Dalla Vecchia C (2014) MNRAS 452:2247. https://doi.org/10.1093/mnras/stu482, http://adsabs.harvard.edu/abs/2014MNRAS.440.2997V

Vegetti S, Koopmans LVE (2009) MNRAS 400:1583. https://doi.org/10.1111/j.1365-2966.2009.15559.x, http://adsabs.harvard.edu/abs/2009MNRAS.400.1583V

Venumadhav T, Cyr-Racine F-Y, Abazajian KN, Hirata CM (2016) Phys Rev D 94:043515. https://doi.org/10.1103/PhysRevD.94.043515, http://adsabs.harvard.edu/abs/2016PhRvD..94d3515V

Viel M, Lesgourgues J, Haehnelt MG, Matarrese S, Riotto A (2005) Phys Rev D 71:063534. https://doi.org/10.1103/PhysRevD.71.063534, http://adsabs.harvard.edu/abs/2005PhRvD..71f3534V

Viel M, Markovič K, Baldi M, Weller J (2012) MNRAS 421:50. https://doi.org/10.1111/j.1365-2966.2011.19910.x, http://adsabs.harvard.edu/abs/2012MNRAS.421...50V

Viel M, Becker GD, Bolton JS, Haehnelt MG (2013) Phys Rev D 88:043502. https://doi.org/10.1103/PhysRevD.88.043502, http://adsabs.harvard.edu/abs/2013PhRvD..88d3502V

Vitvitska M, Klypin AA, Kravtsov AV, Wechsler RH, Primack JR, Bullock JS (2002) ApJ 581:799. https://doi.org/10.1086/344361, http://adsabs.harvard.edu/abs/2002ApJ...581..799V

Wang J, White SDM (2007) MNRAS 380:93. https://doi.org/10.1111/j.1365-2966.2007.12053.x, http://adsabs.harvard.edu/abs/2007MNRAS.380...93W

Wang J et al (2011) MNRAS 413:1373. https://doi.org/10.1111/j.1365-2966.2011.18220.x, http://adsabs.harvard.edu/abs/2011MNRAS.413.1373W

Warren MS, Quinn PJ, Salmon JK, Zurek WH (1992) ApJ 399:405. https://doi.org/10.1086/171937, http://adsabs.harvard.edu/abs/1992ApJ...399..405W

Wechsler RH, Bullock JS, Primack JR, Kravtsov AV, Dekel A (2002) Astrophys J 568:52. https://doi.org/10.1086/338765, http://ads.nao.ac.jp/abs/2002ApJ...568...52W

White SDM (1984) ApJ 286:38. https://doi.org/10.1086/162573, http://adsabs.harvard.edu/abs/1984ApJ...286...38W

Zehavi I et al (2002) ApJ 571:172. https://doi.org/10.1086/339893, http://adsabs.harvard.edu/abs/2002ApJ...571..172Z

Chapter 3
Substructure and Galaxy Formation in Warm Dark Matter Simulations

3.1 Introduction

Non-linear structure formation in thermally produced WDM cosmologies has been extensively studied using simulations in the past few years (e.g. Colín et al. 2000, Bode et al. 2001, Viel et al. 2005, Knebe et al. 2008, Schneider et al. 2012, Lovell et al. 2012, Macciò et al. 2013, Lovell et al. 2014, Reed et al. 2015, Colín et al. 2015, Yang et al. 2015, Bose et al. 2016, Horiuchi et al. 2016).[1] In this chapter we use the *Copernicus Complexio* (COCO- WARM) high resolution N-body simulation to investigate the properties of subhaloes in a WDM model. The observed clumpiness of the Lyman-α forest sets a lower limit to the mass of a dominant thermally produced WDM particle of $m_{\mathrm{WDM}} \geq 3.3$ keV at 95% confidence (Viel et al. 2013); this is consistent with a lower limit set by the observed abundance of satellites in the Milky Way (Kennedy et al. 2014; Lovell et al. 2016). The lower limit to the mass of thermal WDM was increased to $m_{\mathrm{WDM}} \geq 4.35$ keV (95% confidence) by Baur et al. (2016), who repeated the analysis of Viel et al. (2013) with an updated sample of QSO spectra from SDSS-III. These limits, however, depend on uncertain assumptions for thermal history for the intergalactic medium (Garzilli et al. 2015). In our work, as described in Chap. 2, the initial power spectrum was chosen to correspond to a thermal 3.3 keV WDM model. This turns out to have been a fortuitous choice since this power spectrum is very similar to that of the coldest 7 keV sterile neutrino, so constraints on this model can be readily extended to all sterile neutrino models with a 7 keV particle mass.

The formation times, mass functions, spins, shapes, mass profiles and concentrations of haloes in the COCO simulations were presented in Bose et al. (2016) (Chap. 2). Here we focus on the properties of halo substructures. Our simulations are numerically converged down to a halo peak circular velocity of $V_{\mathrm{max}}, \geq 10\,\mathrm{kms}^{-1}$, thus

---

[1]The content of this chapter is based on the article Bose et al. 'Substructure and galaxy formation in the Copernicus Complexio warm dark matter simulations', Monthly Notices of the Royal Astronomical Society, Volume 464, Issue 4, pp. 4520–4533, published 19 October 2016. Reproduced with permission. All rights reserved, https://doi.org/10.1093/mnras/stw2686.

© Springer International Publishing AG, part of Springer Nature 2018
S. Bose, *Beyond ΛCDM* , Springer Theses,
https://doi.org/10.1007/978-3-319-96761-5_3

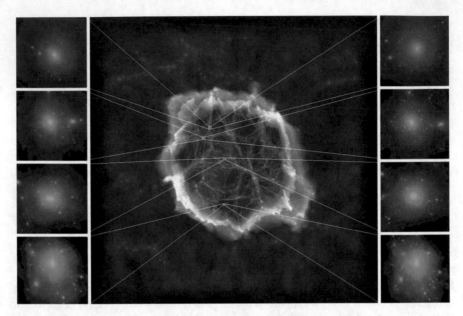

Fig. 3.1 Projected density map in a slice of dimensions $(70.4 \times 70.4 \times 1.5)$ h^{-1} Mpc centred on the COCO high resolution region at $z = 0$. The intensity of the image scales with the number density of particles in the region. The side panels show zooms of a sample of haloes identified at $z = 0$, matched between COCO-WARM (left) and COCO-COLD (right). Figure reproduced from Bose et al. (2017)

allowing statistically meaningful studies of the satellites of the Milky Way. Furthermore, the high resolution of our simulations makes it possible to construct accurate merger trees of even such small haloes and, as a result, we can calculate their observable properties, using the Durham semi-analytical galaxy formation model, GALFORM (Cole et al. 2000; Lacey et al. 2016), a flexible and effective method to implement the best current understanding of galaxy formation physics into an N-body simulation.

The layout of this chapter is as follows. In Sect. 3.2 we investigate the main properties of subhaloes: their population statistics, distribution and internal structure. In Sect. 3.3 we describe the GALFORM model and the modifications required for the WDM case, and compare to predictions for the CDM case. Finally, we summarise our results in Sect. 3.4. A projected density map of the COCO volume at $z = 0$ is shown in Fig. 3.1.

3.2 Dark Matter Substructure

In this section we study the dark matter substructure in the COCO-COLD and COCO-WARM simulations, quantifying their abundance, distribution and internal structure. The general trend we find is that the largest subhaloes in COCO-WARM and COCO-

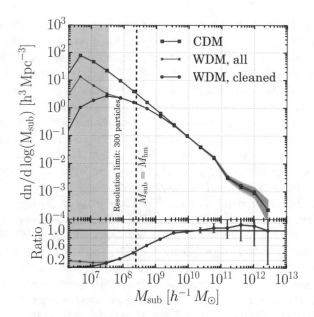

Fig. 3.2 Upper panel: the number density of subhaloes as a function of subhalo mass, M_{sub}, for COCO-COLD (blue), COCO-WARM with all objects included (green), and COCO-WARM with spurious structures removed (red). The shaded region around each curve represents the Poisson uncertainty in the number counts in that bin. The vertical black dashed line marks the half mode mass, M_{hm}, for the 3.3 keV thermal relic. The grey shaded region demarcates the resolution limit of our simulations, set at 300 particles, which was determined by requiring convergence of the mass function compared with the lower-resolution version of COCO-COLD, COLOR- COLD. Lower panel: the ratio of the two COCO-WARM mass functions to the COCO-COLD mass function. Figure reproduced from Bose et al. (2017)

COLD are indistinguishable but differences become increasingly significant below $\sim 5 \times 10^9 \, h^{-1} \, M_\odot$.

3.2.1 The Abundance of Subhaloes

Figure 3.2 shows the present-day differential mass function of subhaloes, dn/d $\log(M_{sub})$, as a function of mass, M_{sub}, in COCO-COLD (blue) and COCO-WARM before (green) and after (red) the removal of artefacts. The lower panel shows the ratio of abundances in COCO-WARM relative to COCO-COLD. The mass function, dn/d $\log(M_{sub})$, is normalised by noting that the irregular volume of the high resolution region has a mean density roughly equal to the mean matter density in the Universe. Combining this with the total mass represented by high resolution particles, we can estimate the volume of the high resolution region.

For $M_{sub} > 10^{10}\,h^{-1}\,M_\odot$ the three mass functions agree very well. These haloes have masses well above the free streaming scale and no spurious objects form on these scales. Below $M_{sub} \sim 5 \times 10^9\,h^{-1}\,M_\odot$, the COCO-WARM mass function begins to peel off from COCO-COLD and by $\sim 3 \times 10^8\,h^{-1}\,M_\odot$ it is suppressed by a factor of two. This mass is close to the "half-mode mass" defined in Eq. 2.3.6, which, in the case of a 3.3 keV thermal relic, has a value: $M_{hm} \approx 2.5 \times 10^8\,h^{-1}\,M_\odot$.

Figure 3.2 shows that the abundance of subhaloes in COCO-WARM is suppressed by a factor of three at M_{hm}. Spurious subhaloes begin to dominate the mass function at a mass an order of magnitude below M_{hm}. Before that happens, and still well above the resolution limit, at $M_{sub} \sim 10^8\,h^{-1}\,M_\odot$, the "cleaned" COCO-WARM mass function (i.e. after subtraction of spurious objects) is already a factor of 5 below the CDM case and shows a sharp turnover. The lower panel in the figure shows these trends more clearly. Removal of the spurious subhaloes is clearly important to obtain an accurate prediction for the abundance of low-mass galaxies in WDM models.

The statistics in COCO are good enough to allow the subhalo mass function to be calculated for different parent (host) halo masses. The result is shown in Fig. 3.3, which gives the (stacked) differential mass functions of subhaloes as a function of the relative mass, $\mu \equiv M_{sub}/M_{200}$ (i.e., the subhalo mass in units of the parent halo mass), in three bins of host halo mass. The COCO-COLD functions are shown with solid lines and the COCO-WARM ones with dashed lines. In both cases, the lines become thinner for subhaloes with fewer than 300 particles. The lower panel of Fig. 3.3 shows the ratio of the differential subhalo mass functions in COCO-WARM to those in COCO-COLD.

The solid lines in the upper panel of Fig. 3.3 illustrate the invariance of the CDM subhalo mass function, when expressed in terms of μ, previously seen by Springel et al. (2008), Gao et al. (2012) and Cautun et al. (2014). The relation is well described by a nearly universal power law (the turnover in the mass function towards low masses is due to incompleteness caused by the resolution of the simulations.) The scale invariance is broken in the case of COCO-WARM, where the mass function deviates from a power law at larger values of μ for smaller host haloes. This can be understood from the fact that, when expressed in units of the host halo mass, the cutoff scale (or, equivalently, M_{hm}) is reached earlier in lower host masses. The abundance of subhaloes is only slightly affected for a host of mass $M_{200} = 10^{13}\,h^{-1}\,M_\odot$, but is strongly suppressed for $M_{200} = 10^{11}\,h^{-1}\,M_\odot$ (for which $\mu = 10^{-3}$ corresponds to $M_{sub} = 10^8\,h^{-1}\,M_\odot$).

Given the ambiguity in the definition of subhalo mass, an alternative property used to count bound substructures is in terms its value of V_{max}, defined as the maximum of the circular velocity curve. Furthermore, this quantity is measurable for many real satellites (where the rotation curve of the satellite can be measured) so it provides a better way than the mass to compare the simulations to observations. The upper panel of Fig. 3.4 shows the "V_{max} function," that is the number of subhaloes as a function of $\nu \equiv V_{max}/V_{200}$, where V_{200} is the circular velocity of the parent halo at r_{200}. Springel et al. (2008) found that the convergence of the V_{max} function improves markedly when V_{max} is corrected for the effects of gravitational softening:

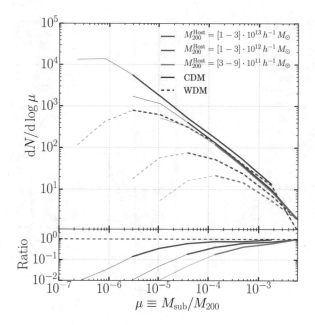

Fig. 3.3 Upper panel: the stacked differential subhalo mass function as a function of parent halo mass, expressed in units of M_{sub}/M_{200}. The CDM case is shown with solid lines and the WDM case with dashed lines. The different colours correspond to different host halo mass ranges as indicated in the legend. The lines become thinner when a given subhalo has fewer than 300 particles i.e., when $\mu \times M^{\mathrm{host}}_{200,\mathrm{mid}} > 300 m_p$, where $M^{\mathrm{host}}_{200,\mathrm{mid}}$ is the centre of the host halo mass bin, and m_p is the high resolution particle mass. Lower panel: ratio of the differential subhalo mass functions in WDM to those in CDM. Figure reproduced from Bose et al. (2017)

$$V^{\mathrm{corr}}_{\mathrm{max}} = V_{\mathrm{max}} \left[1 + (\varepsilon/r_{\mathrm{max}})^2 \right]^{1/2}. \qquad (3.2.1)$$

This correction is important for subhaloes whose r_{max} (the radius at which V_{max} occurs) is not much larger than the gravitational softening, ε. The gravitational softening adopted in COCO ($\varepsilon = 230$ h^{-1} pc) is quite small and we have checked that the correction does not have a significant impact on our results. For CDM, the scale invariance of the subhalo abundance expressed in terms V_{max} is much clearer than when the abundance is expressed in terms of mass, as may be seen by comparing Figs. 3.3 and 3.4, confirming the earlier results of Moore et al. (1999), Kravtsov et al. (2004), Zheng et al. (2005), Springel et al. (2008), Weinberg et al. (2008), Klypin et al. (2011), Wang et al. (2012), Cautun et al. (2014).

It is clear from Figs. 3.3 and 3.4 that, when expressed in dimensionless units such as μ or ν, the subhalo abundance in CDM is close to universal, independent of parent halo mass. In WDM the cutoff in the power spectrum breaks this approximately self-similar behaviour and the subhalo abundance is no longer a universal function.

Fig. 3.4 As Fig. 3.3, but with subhalo abundance expressed as a function of V_{max}^{corr}/V_{200}, where V_{max}^{corr} is the maximum circular velocity, V_{max}, corrected for the effects of gravitational softening as indicated in the legend (see main text). The lines become thinner when $V_{max} < 10\,\mathrm{km s^{-1}}$, which is the circular velocity to which the simulations are complete. Figure reproduced from Bose et al. (2017)

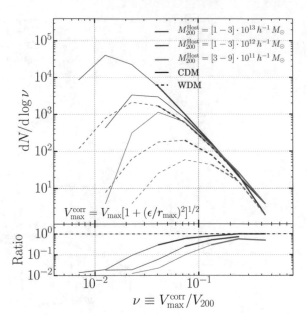

3.2.2 Radial Distribution

Perhaps surprisingly, Springel et al. (2008) found that the normalised radial number density distribution of subhaloes is approximately independent of subhalo mass (see also Ludlow et al. 2009, Hellwing et al. 2016). Han et al. (2016) has provided a simple analytical model that explains this feature, as well as the shape of the subhalo mass function in CDM, as resulting from tidal stripping. The subhalo radial distributions in COCO are shown in Fig. 3.5, which gives the radial number density of subhaloes in different mass ranges, normalised by the mean number density of subhaloes within r_{50} at $z = 0$. The distributions are averaged over 6 parent haloes with mass in the range $1 \times 10^{13}\,h^{-1}\,M_\odot < M_{50}^{Host} < 3 \times 10^{13}\,h^{-1}\,M_\odot$, which are the best resolved in the simulation. The radial positions of the subhaloes are given in units of r_{50}. Only subhaloes resolved with more than 300 particles are included.

The dashed black lines in Fig. 3.5 give a fit to the CDM subhalo number density profiles using the Einasto profile (Einasto 1965; Navarro et al. 2004):

$$\ln\left(\frac{n}{n_{-2}}\right) = -\frac{2}{\alpha}\left[\left(\frac{r}{r_{-2}}\right)^\alpha - 1\right], \qquad (3.2.2)$$

where n_{-2} is the characteristic number density at the scale radius $r = r_{-2}$. The values of r_{-2} and shape parameter, α, given in the legend. The fit is to COCO-COLD profile and the same curve is reproduced in the COCO-WARM panel.

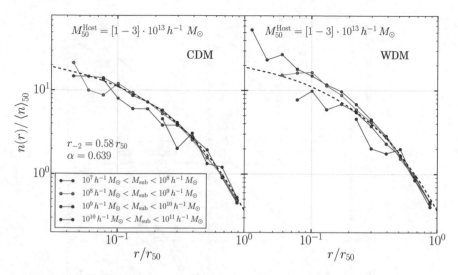

Fig. 3.5 Stacked radial number density profiles of subhaloes, $n(r)$, in different mass ranges (different colours), normalised to the mean number density in that mass range within r_{50} ($\langle n \rangle_{50}$). The profiles are plotted as a function of the distance from the host halo centre (with mass $M_{50}^{\text{Host}} = [1, 2, 3] \cdot 10^{13}\, h^{-1}\, M_{\odot}$). Left: CDM; right: WDM. The dashed black line shows the Einasto profile fit to the COCO-COLD profiles, with the fit parameters r_{-2} and α quoted in the plot. Only subhaloes with more than 300 particles are shown. Figure reproduced from Bose et al. (2017)

The fit to the CDM subhalo profile also provides an excellent fit to the WDM profile, particularly at large radii. There are, however, differences of detail. The distribution of the more massive ($M_{\text{sub}} > 10^9\, h^{-1}\, M_{\odot}$) subhaloes beyond $r > 0.2 r_{50}$ is very similar in COCO-COLD and COCO-WARM. This regime is unaffected by the free streaming cutoff in COCO-WARM. Differences in the radial distribution of these more massive subhaloes can be attributed to small statistics: only six $\sim 10^{13}\, h^{-1}\, M_{\odot}$ haloes contribute to the average shown in Fig. 3.5. The profiles of the less massive subhaloes ($M_{\text{sub}} < 10^9\, h^{-1}\, M_{\odot}$) in WDM are somewhat steeper towards the centre than those in CDM. These subhaloes have masses below the cutoff scale, M_{hm}, and their properties are affected by the cutoff. In particular, they form later than their CDM counterparts of the same mass today and, as a result, they have lower concentrations. These subhaloes experience more mass loss from tidal stripping after infall.

The approximate agreement of the subhalo radial distributions in COCO-COLD and COCO-WARM as well as the differences of detail are consistent with the analytic model proposed by Han et al. (2016). In this model, the $z = 0$ radial number density of subhaloes, n, with mass, m, at distance, R, from the host halo centre is given by:

$$\frac{dn(m, R)}{d \ln m} \propto m^{-\alpha} R^{\gamma} \rho(R) , \tag{3.2.3}$$

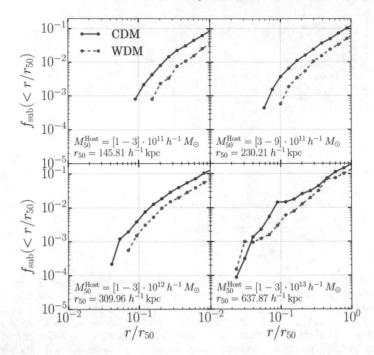

Fig. 3.6 The mass fraction in substructures as a function of dimensionless radial distance from the halo centre, r/r_{50}, for COCO-COLD (solid blue) and COCO-WARM (dashed red) at $z = 0$. The four different panels show results for stacks of host haloes of different mass as indicated in the legend. Only subhaloes with more than 300 particles are included. The value of r_{50} quoted in each panel is the mean over all haloes in each (COCO-COLD) mass bin (the values are similar for COCO-WARM). Figure reproduced from Bose et al. (2017)

where α is the slope of the subhalo mass function evaluated at m, $\rho(R)$ is the density profile of the host dark matter halo, $\gamma = \alpha\beta$, and $\beta \sim 1$ for an NFW density profile. The subhalo number density profile is suppressed relative to the host density profile by the factor R^{γ}. In COCO-COLD, the subhalo mass function follows a single power law but, in COCO-WARM, it has the same slope as in COCO-COLD for $M_{\rm sub} \geq 10^{10}\,h^{-1}\,M_{\odot}$ and a shallower slope below that (see Fig. 3.2). A shallower slope results in a smaller value of α and therefore γ. Equation 3.2.3 then predicts that, compared to CDM, the radial number density profile of small mass subhaloes should be suppressed less relative to the halo density profile for subhaloes. This explains why the two lowest subhalo mass bins in Fig. 3.5 exhibit steeper radial density profiles than the two highest mass bins.

An alternative way to examine the spatial distribution of substructures is to plot the fraction of mass within a given radius that is contained in substructures. This is shown in Fig. 3.6 for different ranges of host halo mass in COCO-COLD and COCO-WARM. The radial distributions have roughly the same shape in the two cases but the subhalo mass fractions are systematically lower in COCO-WARM than in COCO-COLD. In both cases, the substructure mass fractions are higher in more massive host haloes, particularly in the inner regions. For example, for host haloes of mass $M_{50}^{\text{Host}} = (1 - 3) \times 10^{13} \, h^{-1} \, M_\odot$ resolved substructures in COCO-WARM contain about 10% of the halo mass within $r = r_{50}$, but only about 4% for host haloes of mass $M_{50}^{\text{Host}} = (1 - 3) \times 10^{11} \, h^{-1} \, M_\odot$. For reference, haloes (and subhaloes) contain 48% of the total mass in the simulation in COCO-WARM and 56% in COCO-COLD. In CDM simulations these fractions depend on resolution, but not so in COCO-WARM where the cutoff in the power spectrum is resolved.

3.2.3   Internal Structure

The density profiles of WDM haloes and subhaloes are cuspy and well described by the NFW (Navarro et al. 1997) form Lovell et al. (2012), Schneider et al. (2012). However, the later formation times of WDM haloes of mass near the cutoff scale, compared to their CDM counterparts of the same mass, causes them to be less concentrated. In Bose et al. (2016) we characterised the density and mass profiles of haloes in COCO-WARM over a range of halo masses and obtained the concentration-mass relation for WDM haloes (see also Ludlow et al. 2016). In summary, the density profiles of the largest haloes in COCO-WARM (roughly two orders of magnitude above M_{hm}) are indistinguishable from their matched haloes in COCO-COLD, but the profiles of haloes of mass $M_{200} < 7 \times 10^{10} \, h^{-1} \, M_\odot$ have systematically lower concentrations. In contrast with the power-law concentration-mass relation in CDM, the relation in WDM turns over below $\sim 10^{10} \, h^{-1} \, M_\odot$.

Fig. 3.7 Ratio of the infall ($V_{\text{max}}^{\text{inf}}$) to present day ($V_{\text{max}}^{z=0}$) circular velocity, as a function of the present-day circular velocity. The results shown are for 6 stacked host haloes in the mass range $M_{50}^{\text{Host}} = [1, 2, 3] \cdot 10^{13} \, h^{-1} \, M_\odot$, using all subhaloes with more than 300 particles, located within r_{50} of the host centre at $z = 0$. The results for COCO-COLD are shown in blue and for COCO-WARM in red. Figure reproduced from Bose et al. (2017)

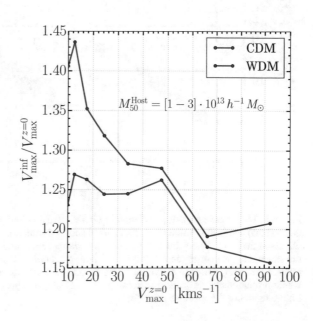

Calculating the concentration of subhaloes from their density profiles is not straightforward because the mass of a subhalo and therefore its "edge" are ambiguous. As Springel et al. (2008) showed, the size calculated by the SUBFIND algorithm (that is the radius of the saddle point in the density profile) coincides with the 'tidal' radius. Defining the concentration of the subhalo using this radius is not particularly useful because its value varies along the orbit. A more useful measure of subhalo concentration is the ratio V_{max}/r_{max}. In both WDM and CDM, the relation between V_{max} and r_{max} has a lower normalisation for subhaloes than for "field haloes" because of tidal stripping.

The fractional change in V_{max} between the moment of infall and the present day is shown in Fig. 3.7 for subhaloes (within r_{50}) of the most massive haloes in COCO-COLD and COCO-WARM, as a function of the present day maximum circular velocity, $V_{max}^{z=0}$ (see also Diemand et al. 2007, Peñarrubia et al. 2008). The largest subhaloes, with $V_{max}^{z=0} \geq 50$ kms^{-1}, experience a reduction in V_{max} by a factor of $1.25 - 1.30$ after infall in both COCO-COLD and COCO-WARM. Subhaloes of lower mass show significant differences between the two simulations. For example, at

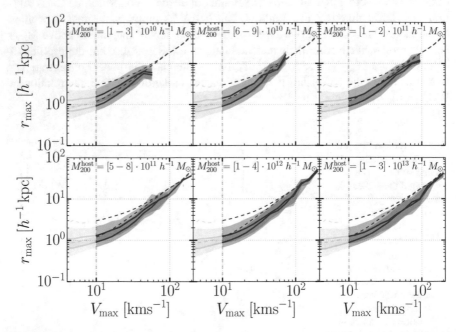

Fig. 3.8 The subhalo r_{max}-V_{max} relation in bins of parent halo mass (different panels) for COCO-COLD (blue) and COCO-WARM (red). Each panel shows results from stacking all host haloes within the given mass bin. The solid line in each case shows the median relation in the stack, whereas the shaded regions correspond to the 16 and 84th percentiles. The dashed lines show the median relation for "field" haloes in each case. The plots are made translucent for $V_{max} < 10$ kms^{-1}, below which resolution effects become increasingly important (see Appendix A in Hellwing et al. 2016). Figure reproduced from Bose et al. (2017)

$V_{max}^{z=0} = 20$ kms^{-1}, COCO-WARM subhaloes have experienced a reduction in V_{max} by a factor of $\sim$1.35 since infall, compared to $\sim$1.25 for COCO-COLD subhaloes.

The $r_{max} - V_{max}$ relations in COCO-COLD and COCO-WARM are shown in Fig. 3.8. For large subhaloes the two are very similar but the relations begin to diverge at values of V_{max} below of a few tens of kilometres per second, depending on the mass of the host halo. In this regime, haloes of a given V_{max} have larger r_{max} in COCO- WARM than in COCO-COLD and are therefore less concentrated. In both COCO-COLD and COCO-WARM subhaloes are more concentrated than field haloes, as a result of tidal stripping, but the difference between field haloes and subhaloes is larger in COCO-WARM than in COCO-COLD. This reflects the greater tidal stripping experienced by COCO-WARM subhaloes, which have lower concentrations when they fall into the host halo. As a result, the concentrations of subhaloes in COCO-WARM increase more than those in COCO-COLD after infall. Overall, however, COCO-WARM subhaloes of a given mass (or V_{max}) still have lower concentrations than COCO-COLD subhaloes. As noted in Hellwing et al. (2016), the importance of tidal stripping depends weakly on host halo mass: at a given V_{max}, the reduction in r_{max} between field haloes and subhaloes is slightly larger for larger host halo masses.

3.3 Galaxy Formation with Warm Dark Matter

Our analysis so far has been restricted to the dark matter properties of a 3.3 keV thermal relic or, equivalently, a 7 keV sterile neutrino with leptogenesis parameter, $L_6 = 8.66$, the "coldest" 7 keV sterile neutrino compatible with the observed 3.5 keV X-ray line. While future gravitational lensing surveys may provide a direct way to measure the mass function of dark matter substructures and thus distinguish CDM from WDM (Vegetti and Koopmans 2009; Li et al. 2016), it is worth investigating whether CDM and WDM can be distinguished with current observations. At high redshift, the observed clumpiness of the Lyman-α forest has been used to rule out WDM models with thermally produced particles of mass $m_{WDM} \leq 3.3$ kev (Viel et al. 2013). As mentioned in Sect. 3.1, constraints obtained from the Lyman-α forest depend on assumptions for the thermal history of the IGM.

To compare the models with other astronomical data we need to populate the dark matter subhaloes with galaxies. This can be done in three ways. One is to use empirical prescriptions such as "abundance matching" (see e.g. Reed et al. 2015) but Sawala et al. (2015) have shown that this technique breaks down for halo masses $<10^{10}$ h^{-1} $M_\odot$ – precisely the scale of interest in WDM. The failure of abundance matching in this regime is due to the physics of reionisation, which inhibits the formation of stars in low mass haloes after the epoch of reionisation, and to the effects of supernovae feedback. A second technique is hydrodynamical simulations but these are computationally expensive and, to date, only limited WDM cosmological simulations have been carried out (e.g. Herpich et al. 2014, Carucci et al. 2015, González-Samaniego et al. 2016). The third approach, the one we use here, is semi-analytical modelling of galaxy formation, a flexible and powerful technique that requires only modest computational resources.

3.3.1 The GALFORM *Semi-analytic Model*

The Durham semi-analytic model of galaxy formation, GALFORM, was introduced by Cole et al. (2000) and has been upgraded regularly as our understanding of the physical processes involved in galaxy formation improves and better observational constraints are obtained. For example, Baugh et al. (2005) introduced a top-heavy IMF in bursts, Bower et al. (2006) introduced AGN feedback and Lagos et al. (2011) introduced a star formation law that depends on the molecular gas content of the ISM. The most recent version of the model Lacey et al. (2016) includes all of these revisions.

We apply the Lacey et al. (2016) version of GALFORM to halo merger trees in COCO-COLD and COCO-WARM. This model includes detailed treatments of gas cooling, star formation, metal production, galaxy mergers and instabilities, black hole growth and feedback from energy released by stellar evolution and AGN. This model was previously used by Kennedy et al. (2014) to set a lower limit to the mass of thermally produced WDM particles.

Details of the modelling in GALFORM may be found in the papers presenting the original formulation of the model (Cole et al. 2000) and its latest version (Lacey et al. 2016). Here we use this latest model for both COCO-COLD and COCO-WARM without any modification.[2]

3.3.2 *Field and Satellite Luminosity Functions*

The galaxy luminosity functions in the b_J and K-bands in COCO-COLD (see also Guo et al. 2015) and COCO-WARM are compared with observational data in Fig. 3.9. The parameters controlling supernova feedback in GALFORM are calibrated to reproduce the observed luminosity functions at $z = 0$ in these bands. The two models predict essentially identical luminosity functions except at faint magnitudes where there are slightly fewer galaxies in WDM, as a result of the lower abundance of small mass haloes in this model. At the faintest magnitudes plotted the difference is only about 25%, smaller than the observational error bars. Due to the small volume of the COCO high resolution region, there are only a few bright galaxies in the simulations, as reflected in the large Poisson errors bars at the brightest magnitudes.

---

[2]Kennedy et al. (2014) found that a small modification to one of the supernovae feedback parameters was required for their WDM models to produce acceptable b_J and K-band luminosity functions at $z = 0$. The particle mass in the model we are considering here, 3.3 keV, is sufficiently large that not even this minor modification is required.

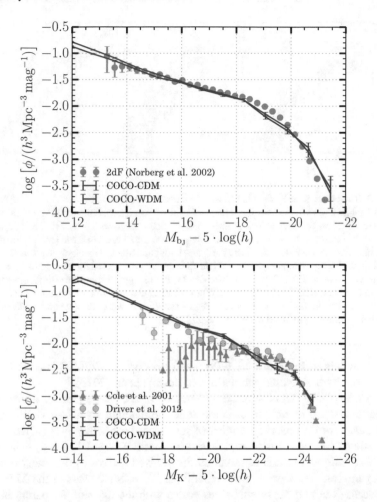

Fig. 3.9 The $z = 0$ b_J- (upper panel) and K-band (lower panel) luminosity functions from GAL-FORM applied to halo merger trees constructed from the COCO-COLD (blue) and COCO-WARM (red) simulations (see text for details). The symbols represent observational data from Norberg et al. (2002), Cole et al. (2001) and Driver et al. (2012). Figure reproduced from Bose et al. (2017)

Fainter galaxies than those plotted in Fig. 3.9 are only detectable in the nearby Universe, particularly in the Local Group. Only a few tens of satellites have been discovered orbiting the haloes of the Milky Way and Andromeda. This number is much smaller than the number of small subhaloes seen in CDM simulations of galactic haloes and this observation has often been used to motivate WDM models. In fact, it has been shown, using a variety of modelling techniques, that most of these small subhaloes are not able to make a visible galaxy either because their gas is heated by reionisation or expelled altogether by supernovae explosions. The earliest explicit demonstration of this simple physics was provided by the semi-

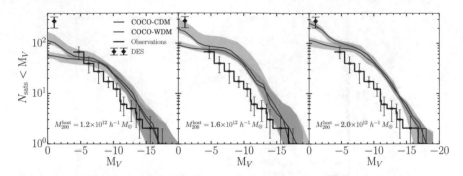

Fig. 3.10 The cumulative V-band luminosity function of satellites within 280 kpc of the centre of Milky Way-like haloes in COCO-COLD (blue) and COCO-WARM (red). Each panel shows the average luminosity function for host haloes in three bins of mass, $M_{200} = 1 - 3 \times 10^{12} \, h^{-1} \, M_\odot$, $1.5 - 1.7 \times 10^{12} \, h^{-1} \, M_\odot$ and $1.8 - 2.1 \times 10^{12} \, h^{-1} \, M_\odot$. The values quoted in the legend are the medians in each bin. The shaded regions indicate the 5 and 95th percentiles. The black step function shows the data for the Milky Way. For $M_V \geq -11$, the data has been corrected for incompleteness and sky coverage by Koposov et al. (2008). For $M_V < -11$, the histogram shows the direct observational data from McConnachie (2012). The black diamond is an extrapolation of the luminosity function to $M_V \sim -1$ after including the ultra-faint dwarf satellites recently discovered by DES (Jethwa et al. 2016). Figure reproduced from Bose et al. (2017)

analytic models of Bullock et al. (2000) and Benson et al. (2002) and the latest by the APOSTLE hydrodynamic simulations of Sawala et al. (2016).

In fact, WDM models can be constrained by the observed number of faint satellites because if the particle mass is too small not enough subhaloes would form to account even for the observed number of satellites in the Milky Way (which may be underestimated because of incompleteness in current surveys). Kennedy et al. (2014) used this argument to set constraints on the allowed masses of thermally produced WDM particles. These constraints depend on the assumed mass of the Milky Way halo because the number of subhaloes scales with the mass of the parent halo (as seen, for example, in Fig. 3.3). Kennedy et al. (2014) find that *all* thermal WDM particle masses are ruled out (at 95% confidence) if the halo of the Milky Way has a mass smaller than $7.7 \times 10^{11} \, h^{-1} \, M_\odot$, while if the mass of the Galactic halo is greater than $1.3 \times 10^{12} \, h^{-1} \, M_\odot$ only WDM particle masses larger than 2 keV are allowed.

We perform a similar analysis here. Figure 3.10 shows the cumulative number of satellites as a function of V-band magnitude, M_V, in COCO-COLD and COCO-WARM for three bins of host halo mass, with median values of 1.2×10^{12}, 1.6×10^{12} and $2.0 \times 10^{12} \, h^{-1} \, M_\odot$. The luminosity function of satellites in the Milky Way, shown by the black solid lines in the figure, include the 11 classical satellites. For $M_V < -11$, the data has been obtained from the direct observations of McConnachie (2012). The abundance of ultra-faint satellites found in the SDSS has been corrected for incompleteness and partial sky coverage by Koposov et al. (2008). The faint objects recently discovered by DES (Bechtol et al. 2015; Drlica-Wagner et al. 2015) are represented by the black diamond following the analysis of Jethwa et al. (2016)

who find that of the 14 newly-detected satellites, 12 have $> 50\%$ probability of having been brought in as satellites of the LMC (at 95% confidence). Jethwa et al. (2016) extrapolate the detected population to estimate that the Milky Way should have ~ 180 satellites within 300 kpc, in addition to 70^{+30}_{-40} Magellanic satellites in the V-band magnitude range $-7 < M_V < -1$ (68% confidence). All observational error bars in Fig. 3.10 are Poisson errors, with volume corrections made where appropriate. In order to match the observational selection, only satellites within 300 kpc of the central galaxy are included.

The satellite luminosity functions are very similar in COCO-COLD and COCO-WARM. Only at magnitudes fainter than $M_V \simeq -4$ does the number of satellites in COCO-WARM begin to drop below the number in COCO-COLD. The models agree with the data so long as the Milky Way halo mass is $M^{host}_{200} \lesssim 1.2 \times 10^{12} \, h^{-1} \, M_\odot$. For $M^{host}_{200} \sim 1.6 \times 10^{12} \, h^{-1} \, M_\odot$, both COCO-COLD and COCO-WARM significantly over-predict the number of satellites even at relatively bright magnitudes, $M_V \sim -10$, where the known sample is unlikely to be significantly incomplete. There is a significant difference in the abundance of satellites with magnitude $M_V \sim -1$, the regime where DES has just begun to uncover ultra-faint dwarf galaxies. These new data could potentially be used to set strong constraints on the mass of the WDM particle. It must be borne in mind that the exact location of this (extrapolated) DES data point depends on the DES selection function, detection efficiency, and assumptions made about isotropy in the distribution of Milky Way satellites. Furthermore, although we have used a well-tested, state-of-the-art model of galaxy formation, these conclusions depend on assumptions in the model, particularly on the treatment of reionisation and supernovae feedback (Hou et al. 2016).

3.3.3 Evolution of the UV Luminosity Function

The evolution of luminosity function in the rest-frame UV traces the star formation history in the Universe. Although still rather scarce and uncertain, data now exist out to redshift $z \sim 10$. Since the formation of structure begins later in WDM models than in CDM we might naïvely expect to find fewer star-forming galaxies at high redshift in COCO-WARM than in COCO-COLD. The actual predictions are shown in Fig. 3.11, which reveals that, in fact, the result is exactly the opposite: at $z > 5$, the UV luminosity function has a higher amplitude in COCO-WARM than in COCO-COLD. The reason for this is that, in CDM, supernovae-driven winds limit the reservoir of cold, potentially star-forming, gas in low-mass galaxies at early times. The brightest UV galaxies at high redshift tend to be starbursts triggered by mergers of these relatively gas poor galaxies (Lacey et al. 2016). By contrast in WDM, the first galaxies that collapse are more massive than their CDM counterparts and more gas rich, thus producing brighter starbursts when they merge. This makes the formation of bright galaxies at high redshift more efficient in WDM than in CDM.

Although both COCO-COLD and COCO-WARM somewhat underpredict current observations at $z > 7$, the data have large statistical, and potentially systematic errors

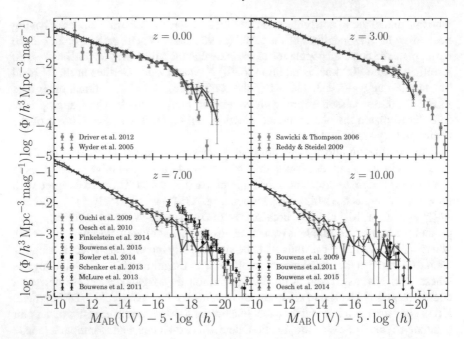

Fig. 3.11 The evolution of the UV luminosity function of all galaxies (centrals and satellites) for $z = 0, 3, 7, 10$. The red lines represents COCO-WARM and the blue COCO-COLD, with Poisson errors plotted. The colour symbols with errorbars show observational data taken from Driver et al. (2012), Wyder et al. (2005), Sawicki and Thompson (2006), Reddy and Steidel (2009), Ouchi et al. (2009), Oesch et al. (2010), Bouwens et al. (2009), Bouwens et al. (2011a, b), Schenker et al. (2013), McLure et al. (2013), Finkelstein et al. (2015), Bowler et al. (2014), Oesch et al. (2014) and Bouwens et al. (2015). Figure reproduced from Bose et al. (2017)

since these objects are rare and current surveys cover relatively small volumes. If anything, COCO-WARM is closer to the data than COCO-COLD. This result is broadly consistent with those of Dayal et al. (2015) who used a simpler model of galaxy formation to derive the UV luminosity function in WDM models. The existence of a population of star-forming galaxies in COCO-WARM at $z > 8$ has the additional benefit that enough ionising photons are produced at early times to reionise the universe by $z \simeq 8$, as required by the optical depth to reionisation inferred from Planck (Planck Collaboration et al. 2014a). Reionisation in WDM models is discussed in detail by Bose et al. (2016).

Figure 3.12 helps visualise the counter-intuitive result just described. In the left panel we plot, as a function of redshift, the stellar mass growth, $M_\star(z)$, averaged over all galaxies with $1 \times 10^7\ h^{-1}\ M_\odot < M_\star < 5 \times 10^7\ h^{-1}\ M_\odot$ at $z = 7$ in COCO-WARM (red) and COCO-COLD (blue). This range of stellar mass corresponds to galaxies brighter than M_{AB} (UV) $\leq\ -17$ in Fig. 3.11. $M_\star(z)$ is normalised to the stellar mass of the galaxy at $z = 7$, $M_\star(z = 7)$. The stellar mass assembly in COCO-WARM is delayed relative to that in COCO-COLD because the earliest progenitors form

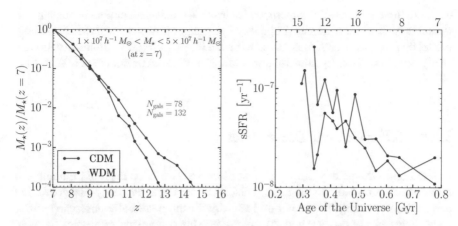

Fig. 3.12 Left panel: the average stellar mass growth of all galaxies with mass $1 \times 10^7 h^{-1} M_\odot < M_\star < 5 \times 10^7 h^{-1} M_\odot$ in COCO-COLD (blue) and COCO-WARM (red). The mass as a function of redshift, $M_\star(z)$, is normalised to the final stellar mass at $z = 7$. The number of galaxies averaged over in each simulation is indicated in the plot with the corresponding colour. Right panel: the specific star formation history as a function of the age of the Universe. The galaxies averaged over are the same as in the left panel. Figure reproduced from Bose et al. (2017)

later in COCO-WARM. For $12 > z > 8$, the build-up of stellar mass is gradual in both COCO-COLD and COCO-WARM, although the slope of the mass growth is steeper in the latter i.e., more stellar mass builds up per unit redshift in COCO-WARM than in COCO-COLD. This is supported by the right panel of Fig. 3.12, which shows the evolution of the specific star formation rate (sSFR) of these galaxies. COCO-WARM galaxies exhibit systematically higher sSFRs than COCO-COLD up to $z = 8$. This is consistent with our earlier suggestion that COCO-WARM galaxies are formed out of more gas-rich progenitors. Mergers of these gas-rich progenitors allows galaxies in COCO-WARM to "catch-up" with those in COCO-COLD after their delayed start of star formation. These findings have been subsequently corroborated by Wang et al. (2017) and Lovell et al. (2018) for different dark matter models to the one assumed in this thesis, demonstrating that the predictions made in this chapter are generic to models with a truncated initial power spectrum.

At $z \leq 3$ the UV luminosity functions in COCO-COLD and COCO-WARM are indistinguishable even down to magnitudes as faint as $M_{AB}(UV) \approx -10$. These galaxies form in haloes of mass $\sim 10^{10} h^{-1} M_\odot$, the scale at which the subhalo mass functions in COCO-WARM just begin to diverge from those in COCO-COLD (see Fig. 3.2). At even fainter magnitudes ($M_{AB}(UV) \geq -7$, not shown), the luminosity function for COCO-WARM is strongly suppressed relative to COCO-COLD but these magnitudes are far below the detection limits of even the JWST.

We have checked that the results in this section are not sensitive to the specific version of the GALFORM model used. The result in Fig. 3.11 holds for the Gonzalez-Perez et al. (2014) model, with and without the assumption of gradual ram-pressure stripping of hot gas in satellite galaxies (Font et al. 2008), as well as for the Hou et al. (2016) model in which supernova feedback is much weaker than in our standard model at high-z and becomes progressively stronger at lower redshift. The simpler model by Dayal et al. (2015) is forced to match the observed UV luminosity function at high-z and cannot, by construction, exhibit any differences between WDM and CDM.

3.3.4   Other Galactic Observables

In addition to the galaxy properties just discussed, we have explored a number of others, such as colour and metallicity distributions; sizes; the Tully-Fisher relation; and spatial clustering. We do not find any significant, potentially observable differences between COCO-COLD and COCO-WARM. This conclusion reinforces the point that, apart from the details discussed in Sects. 3.3.2 and 3.3.3, galaxy formation is very similar in CDM and in a 7 keV sterile neutrino (or a 3.3 keV thermal WDM) model.

3.4   Summary and Discussions

Using the *Copernicus Complexio* (COCO) high resolution dark matter simulations (Hellwing et al. 2016), we have carried out a thorough investigation of the small-scale differences between CDM and a model with the same phases but with a cutoff in the initial power spectrum of fluctuations that can be interpreted either as that of the "coldest" sterile neutrino model compatible with the recently detected 3.5 keV X-ray line or as a 3.3 keV thermal particle model.

The subhalo mass functions in the two models (COCO-COLD and COCO-WARM) are identical at high masses but the number density of COCO-WARM subhaloes begins to fall below that of COCO-COLD subhaloes at $\sim 5 \times 10^9 \, h^{-1} \, M_\odot$ and is very strongly suppressed below $\sim 2.5 \times 10^8 \, h^{-1} \, M_\odot$, the half-mode mass in the initial power spectrum, When the number counts are expressed in units of parent halo properties such as M_{sub}/M_{200} and V_{max}/V_{200}, we find that the subhalo mass and V_{max} functions in COCO-COLD follow a nearly universal profile with little dependence on host halo mass, confirming earlier results (Moore et al. 1999; Kravtsov et al. 2004; Zheng et al. 2005; Springel et al. 2008; Weinberg et al. 2008; Wang et al. 2012; Cautun et al. 2014). This self-similar behaviour does not occur in COCO-WARM.

The normalised radial distribution of subhaloes in both models is independent of the mass of the subhaloes. In the case of COCO-WARM this behaviour extends to the smallest subhaloes in the simulation, with $M_{sub} \simeq 10^8 \, h^{-1} \, M_\odot$, although there is a slight steepening of their profile in the very central parts of the halo. Our findings extend the results from the AQUARIUS and PHOENIX simulations (Springel et al. 2008; Gao et al. 2012) and lend support to the model proposed by Han et al. (2016) in which the mass invariance of the radial distribution results from the effects of tidal stripping. The radial density profiles are well approximated by either the NFW or Einasto forms.

Subhaloes in both COCO-COLD and COCO-WARM are cuspy and follow the NFW form. Small-mass WDM haloes, in general, are less concentrated than CDM haloes of the same mass reflecting their later formation epoch. For WDM subhaloes with $V_{max}^{z=0} \leq 50$ kms^{-1}, the difference is exacerbated because their lower concentrations make them more prone to tidal stripping after they are accreted into the host halo.

In order to check if the two models can be distinguished with current observations, we populated the haloes with model galaxies whose properties were calculated using the Durham semi-analytic galaxy formation model, GALFORM. We used the latest version of GALFORM (Lacey et al. 2016) without needing to adjust any model parameters for COCO-WARM. The COCO-COLD and COCO-WARM b_J and K-band luminosity functions at $z = 0$ are very similar, except at the faintest end where there are slightly fewer dwarfs in COCO-WARM ; both models give a good match to the observations. The same is true at the fainter magnitudes represented by the satellites of the Milky Way: both models agree with current data provided the mass of the Milky Way halo is less than $M_{200} = 1.2 \times 10^{12}$ h^{-1} $M_{\odot}$. The two models could be distinguished if the satellite luminosity function faintwards of $M_V \sim -3$ or -4 could be measured reliably because COCO-WARM predicts about half the number of satellites as COCO-COLD at these luminosities.

The only other significant difference that we have found between COCO-COLD and COCO-WARM is in the UV luminosity function at $z > 7$ where there are more UV-bright galaxies in COCO-WARM than in COCO-COLD. The qualitative difference between the UV luminosity functions in COCO-WARM and COCO-COLD is not strongly affected by the treatment of baryon physics in the GALFORM semi-analytic model. This difference, however (a factor of ~ 2 at $z > 8$), cannot be detected with current data. None of the other galaxy properties we examined: colour and metallicity distributions, scaling relations, spatial clustering, etc. differ in the two models in the regime where these properties can be studied observationally.

In summary, the "coldest" sterile neutrino model compatible with the identification of the recently detected 3.5 keV X-ray line as resulting from the decay of these particles cannot, at present, be distinguished from a CDM model by observations of galaxies, ranging from the satellites of the Milky Way to the brightest starbursts at $z = 10$. The two models are drastically different in their dark matter properties on subgalactic scales where the sterile neutrino model predicts orders of magnitude fewer subhaloes of mass $M \lesssim 10^8$ h^{-1} $M_{\odot}$ than produced in CDM. These small masses are, in principle, accessible to gravitational lensing (Vegetti and Koopmans 2009; Li et al. 2016), and it is to be hoped that future surveys will be able conclusively to rule out one or the other or both of these models.

3.5   New Developments Since Submission of This Thesis in April 2017

One of the main conclusions of the work presented in this chapter is that while CDM and WDM show significant differences in the properties of their dark matter substructures, for WDM particle masses of ~ 3.3 keV, the observed properties of galaxies are almost indistinguishable from CDM (except for the faintest dwarfs or galaxies at high redshift). Observational probes that are directly sensitive to the dark matter substructures themselves offer the most promising means of constraining the mass of the dark matter particle.

An example of such a technique that has gained much interest of late involves the use of stellar streams. More specifically, when such a stream (resulting from the tidal disruption of a globular cluster or dwarf galaxy) encounters subhaloes orbiting within the host halo, the otherwise smooth stream is perturbed, eventually forming a *gap* in the stellar density distribution (e.g. Ibata et al. 2002, Yoon et al. 2011, Carlberg 2012). The distribution of these stream gaps, which grow in size with time, can in principle be used to infer the mass of the subhaloes that create them (potentially sensitive to subhaloes as small as $\sim 10^5$ $M_\odot$). Recent work by Erkal et al. (2016), Banik et al. (2018) have shown how gaps in known streams such as Pal 5 and GD-1 can be used to place nominal constraints on the mass of the dark matter particle. However, this method is not without its complications: gaps in streams can also be created by the interaction of stellar streams with the Galactic bar (Pearson et al. 2017) or with giant molecular clouds (Amorisco et al. 2016). Current and upcoming surveys such as Gaia DR2 and the LSST will uncover many new streams, potentially furthering the constraining power of these methods.

In Li et al. (2016), we explored the power of strong gravitational lensing as a means for detecting low mass substructures. This analysis built on previous work by e.g. Koopmans (2005), Vegetti and Koopmans (2009) who demonstrated that the presence of low mass subhaloes within galaxy groups can distort the surface brightness distribution of Einstein rings around these haloes. Through reconstruction, these distortions can be inverted to get an estimate of the mass of perturbing subhalo.

Figure 3.13 shows the constraints obtained in Li et al. (2016) using mock observations of 50, 100 and 1000 strong lens systems. Here, f_E is the fraction of mass contained in subhaloes at the projected Einstein radius, R_E, and m_c is the 'cutoff' mass that is roughly equivalent to the half-mode mass, M_{hm}, defined in Sect. 3.2.1. Interestingly, this figure shows that even 100 systems are enough to place strong constraints on the cutoff mass at the 2σ level (assuming subhaloes as small as $\sim 10^7$ h^{-1} $M_\odot$ can be detected).

Constraining WDM through strong gravitational lensing is not without its pitfalls. Hydrodynamical simulations performed by D'Onghia et al. (2010), Sawala et al. (2017), Garrison-Kimmel et al. (2017) have shown that even within CDM, a large fraction of low mass subhaloes can be destroyed through interactions with the central disk. The degree of disruption depends on both the mass of the subhalo and its orbital pericentric distance. For example, Sawala et al. (2017) find that at $z = 0$, as much as

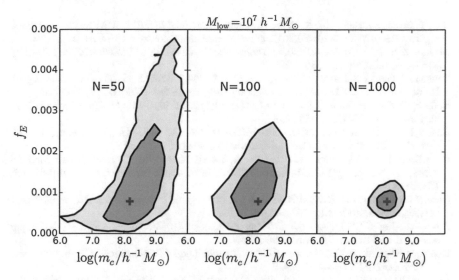

Fig. 3.13 The constraining power on f_E and m_c using 50, 100 and 1000 mock Einstein ring systems for an observation capable of detecting subhaloes as small as $M_{low} = 10^7\,h^{-1}\,M_\odot$. The contours show 68 and 95% confidence levels on this inference. The red crosses show the values of each parameter from the COCO-WARM simulations. Figure reproduced from Li et al. (2016)

50% of the total subhalo population with mass in the range $10^{6.5} - 10^{8.5}\,M_\odot$ can be destroyed within the inner 20 kpc relative to a dark matter only simulation. There is therefore a degeneracy between the absence of low mass substructures which may be induced by the free streaming of dark matter and that which is simply a result baryonic effects.

Fortunately, Li et al. (2017) found that the main contribution to image distortions in Einstein rings come from interloping haloes along the *line of sight*, rather than substructures in the lens. These 'independent' haloes are unaffected by baryonic processes, and their abundance is larger than substructures of the same mass; interloping dark matter haloes therefore boost the differences in the predictions between CDM and WDM. In fact, Li et al. (2017) find that with a detection limit of $10^7\,h^{-1}\,M_\odot$, only 20 strong lens systems would be needed to distinguish between CDM and the 3.3 keV WDM model at 3σ.

References

Amorisco NC, Gómez FA, Vegetti S, White SDM (2016) MNRAS 463:L17. https://doi.org/10.1093/mnrasl/slw148, http://adsabs.harvard.edu/abs/2016MNRAS.463L..17A

Banik N, Bertone G, Bovy J, Bozorgnia N (2018). arXiv:1804.04384

Baugh CM, Lacey CG, Frenk CS, Granato GL, Silva L, Bressan A, Benson AJ, Cole S (2005) MNRAS 356:1191. https://doi.org/10.1111/j.1365-2966.2004.08553.x, http://adsabs.harvard.edu/abs/2005MNRAS.356.1191B

Baur J, Palanque-Delabrouille N, Yèche C, Magneville C, Viel M (2016) JCAP 8:012. https://doi.org/10.1088/1475-7516/2016/08/012, http://adsabs.harvard.edu/abs/2016JCAP...08..012B

Bechtol K et al (2015) ApJ 807:50. https://doi.org/10.1088/0004-637X/807/1/50, http://adsabs.harvard.edu/abs/2015ApJ...807...50B

Benson AJ, Frenk CS, Lacey CG, Baugh CM, Cole S (2002) MNRAS 333:177. https://doi.org/10.1046/j.1365-8711.2002.05388.x, http://adsabs.harvard.edu/abs/2002MNRAS.333..177B

Bode P, Ostriker JP, Turok N (2001) ApJ 556:93. https://doi.org/10.1086/321541, http://adsabs.harvard.edu/abs/2001ApJ...556...93B

Bose S, Frenk CS, Hou J, Lacey CG, Lovell MR (2016) MNRAS 463:3848.https://doi.org/10.1093/mnras/stw2288, http://adsabs.harvard.edu/abs/2016MNRAS.463.3848B

Bose S, Hellwing WA, Frenk CS, Jenkins A, Lovell MR, Helly JC, Li B (2016) MNRAS 455:318.https://doi.org/10.1093/mnras/stv2294,   http://adsabs.harvard.edu/abs/2016MNRAS.455..318B

Bose S, et al (2017) MNRAS 464:4520. https://doi.org/10.1093/mnras/stw2686, http://adsabs.harvard.edu/abs/2017MNRAS.464.4520B

Bouwens RJ et al (2009) ApJ 690:1764.https://doi.org/10.1088/0004-637X/690/2/1764, http://adsabs.harvard.edu/abs/2009ApJ...690.1764B

Bouwens RJ et al (2011a) Nature 469:504. https://doi.org/10.1038/nature09717, http://adsabs.harvard.edu/abs/2011Natur.469..504B

Bouwens RJ et al (2011b) ApJ 737:90. https://doi.org/10.1088/0004-637X/737/2/90, http://adsabs.harvard.edu/abs/2011ApJ...737...90B

Bouwens RJ et al (2015) ApJ 803:34. https://doi.org/10.1088/0004-637X/803/1/34, http://adsabs.harvard.edu/abs/2015ApJ...803...34B

Bower RG, Benson AJ, Malbon R, Helly JC, Frenk CS, Baugh CM, Cole S, Lacey CG (2006) MNRAS 370:645. https://doi.org/10.1111/j.1365-2966.2006.10519.x, http://adsabs.harvard.edu/abs/2006MNRAS.370..645B

Bowler RAA et al (2014) MNRAS 440:2810. https://doi.org/10.1093/mnras/stu449, http://adsabs.harvard.edu/abs/2014MNRAS.440.2810B

Bullock JS, Kravtsov AV, Weinberg DH (2000) ApJ 539:517. https://doi.org/10.1086/309279, http://adsabs.harvard.edu/abs/2000ApJ...539..517B

Carlberg RG (2012) ApJ 748:20. https://doi.org/10.1088/0004-637X/748/1/20, http://adsabs.harvard.edu/abs/2012ApJ...748...20C

Carucci IP, Villaescusa-Navarro F, Viel M, Lapi A (2015) JCAP 7:047. https://doi.org/10.1088/1475-7516/2015/07/047, http://adsabs.harvard.edu/abs/2015JCAP...07..047C

Cautun M, Hellwing WA, van de Weygaert R, Frenk CS, Jones BJT, Sawala T (2014) MNRAS 445:1820. https://doi.org/10.1093/mnras/stu1829, http://adsabs.harvard.edu/abs/2014MNRAS.445.1820C

Cole S et al (2001) MNRAS 326:255. https://doi.org/10.1046/j.1365-8711.2001.04591.x, http://adsabs.harvard.edu/abs/2001MNRAS.326..255C

Cole S, Lacey CG, Baugh CM, Frenk CS (2000) MNRAS 319:168. https://doi.org/10.1046/j.1365-8711.2000.03879.x, http://adsabs.harvard.edu/abs/2000MNRAS.319..168C

Colín P, Avila-Reese V, Valenzuela O (2000) ApJ 542:622. https://doi.org/10.1086/317057, http://adsabs.harvard.edu/abs/2000ApJ...542..622C

Colín P, Avila-Reese V, González-Samaniego A, Velázquez H (2015) ApJ 803:28. https://doi.org/10.1088/0004-637X/803/1/28, http://adsabs.harvard.edu/abs/2015ApJ...803...28C

Dayal P, Mesinger A, Pacucci F (2015) ApJ 806:67. https://doi.org/10.1088/0004-637X/806/1/67, http://adsabs.harvard.edu/abs/2015ApJ...806...67D

Diemand J, Kuhlen M, Madau P (2007) ApJ 667:859. https://doi.org/10.1086/520573, http://adsabs.harvard.edu/abs/2007ApJ...667..859D

D'Onghia E, Springel V, Hernquist L, Keres D (2010) ApJ 709:1138. https://doi.org/10.1088/0004-637X/709/2/1138, http://adsabs.harvard.edu/abs/2010ApJ...709.1138D

Driver SP et al (2012) MNRAS 427:3244. https://doi.org/10.1111/j.1365-2966.2012.22036.x, http://adsabs.harvard.edu/abs/2012MNRAS.427.3244D

Drlica-Wagner A et al (2015) ApJ 813:109. https://doi.org/10.1088/0004-637X/813/2/109, http://adsabs.harvard.edu/abs/2015ApJ...813..109D

Einasto J (1965) Trudy Astrofizicheskogo Instituta Alma-Ata 5:87. http://adsabs.harvard.edu/abs/1965TrAlm...5...87E

Erkal D, Belokurov V, Bovy J, Sanders JL (2016) MNRAS 463:102. https://doi.org/10.1093/mnras/stw1957, http://adsabs.harvard.edu/abs/2016MNRAS.463..102E

Finkelstein SL et al (2015) ApJ 810:71. https://doi.org/10.1088/0004-637X/810/1/71, http://adsabs.harvard.edu/abs/2015ApJ...810...71F

Font AS et al (2008) MNRAS 389:1619. https://doi.org/10.1111/j.1365-2966.2008.13698.x, http://adsabs.harvard.edu/abs/2008MNRAS.389.1619F

Gao L, Navarro JF, Frenk CS, Jenkins A, Springel V, White SDM (2012) MNRAS 425:2169. https://doi.org/10.1111/j.1365-2966.2012.21564.x, http://adsabs.harvard.edu/abs/2012MNRAS.425.2169G

Garrison-Kimmel S. et al (2017) MNRAS 471:1709. https://doi.org/10.1093/mnras/stx1710, http://adsabs.harvard.edu/abs/2017MNRAS.471.1709G

Garzilli A, Boyarsky A, Ruchayskiy O (2015). arXiv:1510.07006

Gonzalez-Perez V, Lacey CG, Baugh CM, Lagos CDP, Helly J, Campbell DJR, Mitchell PD (2014) MNRAS 439:264. https://doi.org/10.1093/mnras/stt2410, http://adsabs.harvard.edu/abs/2014MNRAS.439..264G

González-Samaniego A, Avila-Reese V, Colín P (2016) ApJ 819:101. https://doi.org/10.3847/0004-637X/819/2/101, http://adsabs.harvard.edu/abs/2016ApJ...819..101G

Guo Q, Cooper AP, Frenk C, Helly J, Hellwing WA (2015) MNRAS 454:550. https://doi.org/10.1093/mnras/stv1938, http://adsabs.harvard.edu/abs/2015MNRAS.454..550G

Han J, Cole S, Frenk CS, Jing Y (2016) MNRAS 457:1208. https://doi.org/10.1093/mnras/stv2900, http://adsabs.harvard.edu/abs/2016MNRAS.457.1208H

Hellwing WA, Frenk CS, Cautun M, Bose S, Helly J, Jenkins A, Sawala T, Cytowski M (2016) MNRAS 457:3492. https://doi.org/10.1093/mnras/stw214, http://adsabs.harvard.edu/abs/2016MNRAS.457.3492H

Herpich J, Stinson GS, Macciò AV, Brook C, Wadsley J, Couchman HMP, Quinn T (2014) MNRAS 437:293. https://doi.org/10.1093/mnras/stt1883, http://adsabs.harvard.edu/abs/2014MNRAS.437..293H

Horiuchi S, Bozek B, Abazajian KN, Boylan-Kolchin M, Bullock JS, Garrison-Kimmel S, Onorbe J (2016) MNRAS 456:4346. https://doi.org/10.1093/mnras/stv2922, http://adsabs.harvard.edu/abs/2016MNRAS.456.4346H

Hou J, Frenk CS, Lacey CG, Bose S (2016) MNRAS 463:1224. https://doi.org/10.1093/mnras/stv2033, http://adsabs.harvard.edu/abs/2016MNRAS.463.1224H

Ibata RA, Lewis GF, Irwin MJ, Quinn T (2002) MNRAS 332:915. https://doi.org/10.1046/j.1365-8711.2002.05358.x, http://adsabs.harvard.edu/abs/2002MNRAS.332..915I

Jethwa P, Erkal D, Belokurov V (2016) MNRAS 461:2212. https://doi.org/10.1093/mnras/stw1343, http://adsabs.harvard.edu/abs/2016MNRAS.461.2212J

Kennedy R, Frenk C, Cole S, Benson A (2014) MNRAS 442:2487. https://doi.org/10.1093/mnras/stu719, http://adsabs.harvard.edu/abs/2014MNRAS.442.2487K

Klypin AA, Trujillo-Gomez S, Primack J (2011) ApJ 740:102. https://doi.org/10.1088/0004-637X/740/2/102, http://adsabs.harvard.edu/abs/2011ApJ...740..102K

Knebe A, Arnold B, Power C, Gibson BK (2008) MNRAS 386:1029. https://doi.org/10.1111/j.1365-2966.2008.13102.x, http://adsabs.harvard.edu/abs/2008MNRAS.386.1029K

Koopmans LVE (2005) MNRAS 363:1136. https://doi.org/10.1111/j.1365-2966.2005.09523.x, http://adsabs.harvard.edu/abs/2005MNRAS.363.1136K

Koposov S et al (2008) ApJ 686:279. https://doi.org/10.1086/589911, http://adsabs.harvard.edu/abs/2008ApJ...686..279K

Kravtsov AV, Berlind AA, Wechsler RH, Klypin AA, Gottlöber S, Allgood B, Primack JR (2004) ApJ 609:35. https://doi.org/10.1086/420959, http://adsabs.harvard.edu/abs/2004ApJ...609...35K

Lacey CG et al (2016) MNRAS 462:3854. https://doi.org/10.1093/mnras/stw1888, http://adsabs. harvard.edu/abs/2016MNRAS.462.3854L

Lagos CDP, Lacey CG, Baugh CM, Bower RG, Benson AJ (2011) MNRAS 416:1566. https:// doi.org/10.1111/j.1365-2966.2011.19160.x, http://adsabs.harvard.edu/abs/2011MNRAS.416. 1566L

Li R, Frenk CS, Cole S, Gao L, Bose S, Hellwing WA (2016) MNRAS 460:363. https://doi.org/10. 1093/mnras/stw939, http://adsabs.harvard.edu/abs/2016MNRAS.460..363L

Li R, Frenk CS, Cole S, Wang Q, Gao L (2017) MNRAS 468:1426. https://doi.org/10.1093/mnras/ stx554, http://adsabs.harvard.edu/abs/2017MNRAS.468.1426L

Lovell MR et al (2012) MNRAS 420:2318. https://doi.org/10.1111/j.1365-2966.2011.20200.x, http://adsabs.harvard.edu/abs/2012MNRAS.420.2318L

Lovell MR, Frenk CS, Eke VR, Jenkins A, Gao L, Theuns T (2014) MNRAS 439:300. https://doi. org/10.1093/mnras/stt2431, http://adsabs.harvard.edu/abs/2014MNRAS.439..300L

Lovell MR et al (2016) MNRAS 461:60. https://doi.org/10.1093/mnras/stw1317, http://adsabs. harvard.edu/abs/2016MNRAS.461...60L

Lovell MR et al (2018) MNRAS. https://doi.org/10.1093/mnras/sty818

Ludlow AD, Navarro JF, Springel V, Jenkins A, Frenk CS, Helmi A (2009) ApJ 692:931. https:// doi.org/10.1088/0004-637X/692/1/931, http://adsabs.harvard.edu/abs/2009ApJ...692..931L

Ludlow AD, Bose S, Angulo RE, Wang L, Hellwing WA, Navarro JF, Cole S, Frenk CS (2016) MNRAS 460:1214. https://doi.org/10.1093/mnras/stw1046, http://adsabs.harvard.edu/ abs/2016MNRAS.460.1214L

Macciò AV, Ruchayskiy O, Boyarsky A, Muñoz-Cuartas JC (2013) MNRAS 428:882. https://doi. org/10.1093/mnras/sts078, http://adsabs.harvard.edu/abs/2013MNRAS.428..882M

McConnachie AW (2012) AJ 144:4. https://doi.org/10.1088/0004-6256/144/1/4, http://adsabs. harvard.edu/abs/2012AJ....144....4M

McLure RJ et al (2013) MNRAS 432:2696. https://doi.org/10.1093/mnras/stt627, http://adsabs. harvard.edu/abs/2013MNRAS.432.2696M

Moore B, Ghigna S, Governato F, Lake G, Quinn T, Stadel J, Tozzi P (1999) ApJL 524:L19. https:// doi.org/10.1086/312287, http://adsabs.harvard.edu/abs/1999ApJ...524L..19M

Navarro JF, Frenk CS, White SDM (1997) ApJ 490:493. http://adsabs.harvard.edu/abs/1997ApJ... 490..493N

Navarro JF et al (2004) MNRAS 349:1039.https://doi.org/10.1111/j.1365-2966.2004.07586.x, http://adsabs.harvard.edu/abs/2004MNRAS.349.1039N

Norberg P et al (2002) MNRAS 336:907. https://doi.org/10.1046/j.1365-8711.2002.05831.x, http:// adsabs.harvard.edu/abs/2002MNRAS.336..907N

Oesch PA et al (2010) ApJL 709:L16. https://doi.org/10.1088/2041-8205/709/1/L16, http://adsabs. harvard.edu/abs/2010ApJ...709L..16O

Oesch PA et al (2014) ApJ 786:108.https://doi.org/10.1088/0004-637X/786/2/108, http://adsabs. harvard.edu/abs/2014ApJ...786..108O

Ouchi M et al (2009) ApJ 706:1136. https://doi.org/10.1088/0004-637X/706/2/1136, http://adsabs. harvard.edu/abs/2009ApJ...706.1136O

Pearson S, Price-Whelan AM, Johnston KV (2017) Nat Astron 1:633. https://doi.org/10.1038/ s41550-017-0220-3, http://adsabs.harvard.edu/abs/2017NatAs...1..633P

Peñarrubia J, Navarro JF, McConnachie AW (2008) ApJ 673:226. https://doi.org/10.1086/523686, http://adsabs.harvard.edu/abs/2008ApJ...673..226P

Planck Collaboration et al (2014) A&A. https://doi.org/10.1051/0004-6361/201321529, http:// adsabs.harvard.edu/abs/2014A

Reddy NA, Steidel CC (2009) ApJ 692:778. https://doi.org/10.1088/0004-637X/692/1/778, http:// adsabs.harvard.edu/abs/2009ApJ...692..778R

Reed DS, Schneider A, Smith RE, Potter D, Stadel J, Moore B (2015) MNRAS 451:4413. https:// doi.org/10.1093/mnras/stv1233, http://adsabs.harvard.edu/abs/2015MNRAS.451.4413R

Sawala T et al (2015) MNRAS 448:2941. https://doi.org/10.1093/mnras/stu2753, http://adsabs. harvard.edu/abs/2015MNRAS.448.2941S

Sawala T et al (2016) MNRAS 457:1931. https://doi.org/10.1093/mnras/stw145, http://adsabs.harvard.edu/abs/2016MNRAS.457.1931S

Sawala T, Pihajoki P, Johansson PH, Frenk CS, Navarro JF, Oman KA, White SDM (2017) MNRAS 467:4383. https://doi.org/10.1093/mnras/stx360, http://adsabs.harvard.edu/abs/2017MNRAS.467.4383S

Sawicki M, Thompson D (2006) ApJ 642:653. https://doi.org/10.1086/500999, http://adsabs.harvard.edu/abs/2006ApJ...642..653S

Schenker MA et al (2013) ApJ 768:196. https://doi.org/10.1088/0004-637X/768/2/196, http://adsabs.harvard.edu/abs/2013ApJ...768..196S

Schneider A, Smith RE, Macciò AV, Moore B (2012) MNRAS 424:684. https://doi.org/10.1111/j.1365-2966.2012.21252.x, http://adsabs.harvard.edu/abs/2012MNRAS.424..684S

Springel V et al (2008) MNRAS 391:1685. https://doi.org/10.1111/j.1365-2966.2008.14066.x, http://adsabs.harvard.edu/abs/2008MNRAS.391.1685S

Vegetti S, Koopmans LVE (2009) MNRAS 400:1583. https://doi.org/10.1111/j.1365-2966.2009.15559.x, http://adsabs.harvard.edu/abs/2009MNRAS.400.1583V

Viel M, Lesgourgues J, Haehnelt MG, Matarrese S, Riotto A (2005) Phys Rev D 71:063534. https://doi.org/10.1103/PhysRevD.71.063534, http://adsabs.harvard.edu/abs/2005PhRvD..71f3534V

Viel M, Becker GD, Bolton JS, Haehnelt MG (2013) Phys Rev D 88:043502. https://doi.org/10.1103/PhysRevD.88.043502, http://adsabs.harvard.edu/abs/2013PhRvD..88d3502V

Wang J, Frenk CS, Navarro JF, Gao L, Sawala T (2012) MNRAS 424:2715. https://doi.org/10.1111/j.1365-2966.2012.21357.x, http://adsabs.harvard.edu/abs/2012MNRAS.424.2715W

Wang L et al (2017) MNRAS 468:4579. https://doi.org/10.1093/mnras/stx788, http://adsabs.harvard.edu/abs/2017MNRAS.468.4579W

Weinberg DH, Colombi S, Davé R, Katz N (2008) ApJ 678:6. https://doi.org/10.1086/524646, http://adsabs.harvard.edu/abs/2008ApJ...678....6W

Wyder TK et al (2005) ApJL 619:L15. https://doi.org/10.1086/424735, http://adsabs.harvard.edu/abs/2005ApJ...619L..15W

Yang LF, Neyrinck MC, Aragón-Calvo MA, Falck B, Silk J (2015) MNRAS 451:3606. https://doi.org/10.1093/mnras/stv1087, http://adsabs.harvard.edu/abs/2015MNRAS.451.3606Y

Yoon JH, Johnston KV, Hogg DW (2011) ApJ 731:58. https://doi.org/10.1088/0004-637X/731/1/58, http://adsabs.harvard.edu/abs/2011ApJ...731...58Y

Zheng Z et al (2005) ApJ 633:791. https://doi.org/10.1086/466510, http://adsabs.harvard.edu/abs/2005ApJ...633..791Z

Chapter 4
Reionisation in Sterile Neutrino Cosmologies

4.1 Introduction

From[1] the point of view of cosmology, the defining property of keV mass sterile neutrinos is that they behave as *warm dark matter* (WDM). In contrast to CDM, warm particles are kinematically energetic at early times and thus free stream out of small-scale primordial perturbations, inducing a cut-off in the power spectrum of density fluctuations. On large scales unaffected by the free streaming cut-off, structure formation is very similar in CDM and sterile neutrino cosmologies (and in WDM in general), but on scales comparable to or smaller than the cut-off, structure formation proceeds in a fundamentally different way in the two cases. No haloes form below a certain mass scale determined by the cut-off and the formation of small haloes above the cut-off is delayed (see Colín et al. 2000, Bode et al. 2001, Avila-Reese et al. 2001, Viel et al. 2005, Lovell et al. 2012, Schneider et al. 2012, Bose et al. 2016a, 2017).

For a 7 keV sterile neutrino, the cut-off mass is $\sim 10^9 M_\odot$. Thus, potentially observable differences from CDM would emerge on subgalactic scales and at high redshifts when the delayed onset of structure formation might become apparent. The Local Group and the early Universe are thus good hunting grounds for tell-tale signs that might distinguish warm from cold dark matter. There is now a wealth of observational data for small galaxies in the Local Group (e.g. Koposov et al. 2008, McConnachie 2012), as well as measurements of the abundance of galaxies at high redshifts (e.g. McLure et al. 2013, Bouwens et al. 2015) and estimates of the redshift of reionisation (Planck Collaboration et al. 2016). One might hope that these data could constrain the parameters of WDM models (e.g. Schultz et al. 2014, Abazajian 2014, Calura et al. 2014, Dayal et al. 2017, 2015, Governato et al. 2015, Lovell et al. 2016, Maio and Viel 2015, Bozek et al. 2016).

---

[1]The content of this chapter is based on the article Bose et al. 'Reionization in sterile neutrino cosmologies', Monthly Notices of the Royal Astronomical Society, Volume 463, Issue 4, p. 3848–3859, published 21 December 2016. Reproduced with permission. All rights reserved, https://doi.org/10.1093/mnras/stw2288.

© Springer International Publishing AG, part of Springer Nature 2018
S. Bose, *Beyond ΛCDM* , Springer Theses,
https://doi.org/10.1007/978-3-319-96761-5_4

In this work, we address these questions using the Durham semi-analytic model of galaxy formation, GALFORM (Cole et al. 2000; Lacey et al. 2016), applied both to CDM and sterile neutrino dark matter. The model follows the formation of galaxies in detail using a Monte Carlo technique for calculating halo merger trees and well-tested models for the baryon physics that result in the formation of visible galaxies. GAL-FORM predicts the properties of the galaxy population at all times. This approach has the advantage that it can easily generate large statistical samples of galaxies at high resolution for a variety of dark matter models which would be prohibitive in terms of computational time with the current generation of hydrodynamic simulations.

Here, we are particularly interested in sterile neutrinos that could decay to produce two 3.5 keV photons. We therefore fix the mass $M_1 = 7$ keV. At this mass, the 'warmest' and 'coldest' sterile neutrino models that achieve the correct dark matter density correspond to $L_6 = 700$ and $L_6 = 8$ respectively. By this we mean that the $L_6 = 700$ model exhibits deviations from CDM at larger mass scales than the $L_6 = 8$ model, which produces similar structure to CDM down to the scale of dwarf galaxies.

For the $L_6 = 700$ case, however, the corresponding mixing angle (which we remind the reader is now *fixed*) does not lead to the X-ray decay flux required to account for the observations of Bulbul et al. (2014b), Boyarsky et al. (2014). For this reason, we additionally consider the case $L_6 = 12$, which corresponds to the warmest 7 keV sterile neutrino model that has the correct dark matter abundance *and* produces the correct flux at 3.5 keV. This information is summarised in Table 4.1. Here, we also quote the characteristic "half-mode" wavenumber (c.f. Eqs. 2.3.5 and 2.3.6), k_{hm}. We remind the reader that k_{hm} characterises the 'warmth' of the model. The most extreme case ($L_6 = 700$) has $k_{hm} = 16.05$ h/Mpc, whereas the model closest to CDM ($L_6 = 8$) has $k_{hm} = 44.14$ h/Mpc.

Figure 4.1 shows the linear power spectrum (in arbitrary units) of these three models ($L_6 = (8, 12, 700)$), with the CDM power spectrum also plotted for com-

Table 4.1 Properties of the four dark matter models studied in this chapter: CDM and 7 keV sterile neutrino models with lepton asymmetry, $L_6 = (8, 12, 700)$. The quantity k_{hm} is the wavenumber at which the amplitude of the power spectrum is $1/4$ that of the CDM amplitude; it is a measure of the "warmth" of the model. The last three columns indicate whether the model gives (1) the correct dark matter density; (2) whether the particle can decay to produce a line at 3.5 keV; and (3) whether the corresponding mixing angle can produce an X-ray decay flux consistent with the observations of Boyarsky et al. (2014), Bulbul et al. (2014b). Figure reproduced from Bose et al. (2016b)

Model; L_6	k_{hm} [h/Mpc]	Right DM abundance?	Decay at 3.5 keV?	Flux consistent with 3.5 keV X-ray line?
CDM; –	–	✓	✗	✗
7 keV; 8	44.14	✓	✓	✓
7 keV; 12	23.27	✓	✓	✓
7 keV; 700	16.05	✓	✓	✗

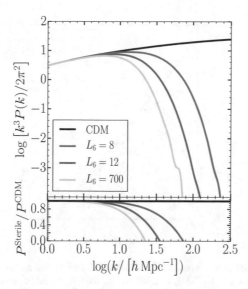

Fig. 4.1 *Top panel*: The dimensionless matter power spectra for the different dark matter candidates considered in this chapter. In addition to CDM, we consider a 7 keV sterile neutrino with three values of $L_6 = (8, 12, 700)$, shown with the colours indicated in the legend. For the same sterile neutrino mass, different L_6 values lead to deviations from CDM on different scales, with the most extreme case being the $L_6 = 700$ model. *Bottom panel*: The ratio of each power spectrum to that of CDM. Figure reproduced from Bose et al. (2016b)

parison. The power spectra for the sterile neutrino models were computed by first calculating the momentum distribution functions for these models using the methods outlined by Laine and Shaposhnikov (2008), Ghiglieri and Laine (2015), and using these to solve the Boltzmann equation with a modified version of the CAMB code (Lewis et al. 2000; Boyarsky et al. 2009a, b; Lovell et al. 2016). The cosmological parameters assumed are those derived from Planck Collaboration et al. (2016): $\Omega_m = 0.307$, $\Omega_\Lambda = 0.693$, $\Omega_b = 0.0483$, $h = 0.678$, $\sigma_8 = 0.823$, and $n_s = 0.961$. The most striking feature is how, for the same 7 keV sterile neutrino, the scale of the cut-off (as measured by the half-mode wavenumber, k) changes with L_6. The cutoff in the $L_6 = 8$ power spectrum occurs at a similar scale to that introduced by a 3.3 keV thermal relic, which, at 95% confidence, is the lower limit on the WDM particle mass set by constraints from the Lyman-α forest (Viel et al. 2013, although see Baur et al. 2016 for a revised lower limit). The $L_6 = 12$ case is therefore in tension with the lower limits from the Lyman-α forest, but it should be noted that the derived lower limits are sensitive to assumptions made for the thermal history of the IGM (Garzilli et al. 2015).

The truncated power spectra in the three sterile neutrino models results in a suppression in the abundance of haloes (and by extension, the galaxies in them) at different mass scales in the different models. This is illustrated in Fig. 4.2 where we show the $z = 0$ halo mass functions for CDM and for $L_6 = (8, 12, 700)$, as predicted

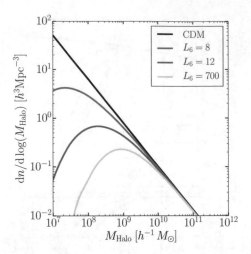

Fig. 4.2 The $z = 0$ halo mass functions for CDM and 7 keV sterile neutrino models with leptogenesis parameter, $L_6 = (8, 12, 700)$, as predicted by the ellipsoidal collapse model of Sheth and Tormen (1999), calculated using Eqs. 4.1.1 and 4.1.2. The different cut-off scales for the sterile neutrino power spectra in Fig. 4.1 are reflected in the different mass scales at which the corresponding halo mass functions are suppressed below the CDM mass function. Figure reproduced from Bose et al. (2016b)

by the ellipsoidal collapse formalism of Sheth and Tormen (1999). In this model, the number density of haloes within a logarithmic interval in mass (dn/d log M_{halo}) is quantified by:

$$\frac{dn}{d \log M_{halo}} = \frac{\bar{\rho}}{M_{halo}} f(\nu) \left| \frac{d \log \sigma^{-1}}{d \log M_{halo}} \right|, \qquad (4.1.1)$$

where $\bar{\rho}$ is the mean matter density of the Universe, $\nu = \delta_c / \sigma(M_{halo})$, $\delta_c = 1.686$ is the density threshold required for collapse and $\sigma(M_{halo})$ is the variance of the density field, smoothed at a scale, M_{halo} (see Sect. 4.2.3). In the ellipsoidal collapse model the multiplicity function, $f(\nu)$, takes the form:

$$f(\nu) = A \sqrt{\frac{2q\nu}{\pi}} \left[1 + (q\nu)^{-p} \right] e^{-q\nu/2}, \qquad (4.1.2)$$

where $A = 0.3222$, $q = 0.707$ and $p = 0.3$. Figure 4.2 shows how the mass functions in the sterile neutrino models peel off from CDM at different mass scales directly related to k_{hm}. The halo masses corresponding to these wavenumbers can be estimated by:

$$M_{hm} = \frac{4}{3} \pi \bar{\rho} \left(\frac{\pi}{k_{hm}} \right)^3, \qquad (4.1.3)$$

giving $M_{hm} = (1.1 \times 10^8, 7.8 \times 10^8, 2.3 \times 10^9)\ h^{-1}\ M_\odot$ for $L_6 = (8, 12, 700)$ respectively. Clearly, the largest suppression in halo abundance relative to CDM occurs for the $L_6 = 700$ case, and the least for the $L_6 = 8$ case, consistent with our discussion of the significance of the characteristic scale k_{hm}. For example, at $z = 0$, there are half as many $\sim 10^8\ h^{-1}\ M_\odot$ in $L_6 = 8$ as in CDM. By comparison, there are ~ 150 times fewer haloes at the same mass scale for $L_6 = 700$ relative to CDM. The $L_6 = 12$ model lies in between these two cases, producing ~ 20 times fewer haloes of $10^8\ h^{-1}\ M_\odot$.

The rest of this chapter is structured as follows. In Sect. 4.2 we describe the astrophysical motivation behind this work, as well as the semi-analytic model, GALFORM, used in our analysis. Our results are presented in Sect. 4.3 and our main conclusions summarised in Sect. 4.4.

4.2 Galaxy Formation

We begin by discussing the astrophysical quantities and observables that we will use to constrain sterile neutrino models. We then briefly introduce the specific implementation of GALFORM that we will use to predict these quantities for both CDM and sterile neutrino models. We build upon the ideas and methods laid out by Hou et al. (2016, hereafter Hou15).

4.2.1 A Galactic "Tug-of-War"

One of the most important physical processes involved in galaxy formation is supernova feedback (SNfb). By ejecting cold gas from galaxies, SNfb regulates star formation, inhibiting galaxy formation in small mass haloes (Larson 1974; White and Frenk 1991). SNfb is thought to be responsible for the relatively flat galaxy stellar mass and luminosity functions compared to the steeply rising halo mass function predicted by N-body simulations for ΛCDM (e.g. Jenkins et al. 2001; Tinker et al. 2008; Kauffmann et al. 1993; Cole et al. 1994). On the smallest scales, SNfb, in conjunction with photoionisation of gas in the early Universe, can explain the small number of faint satellite galaxies seen around galaxies like the Milky Way in this model (Efstathiou 1992; Benson et al. 2003; Sawala et al. 2015).

Unless AGN contribute a significant number of ionising photons (Madau and Haardt 2015; Khaire et al. 2016), SNfb cannot be so strong as to suppress the production of ionising photons at high redshift required to reionise the Universe by $z \sim 6$, as inferred from QSO absorption lines (Mitra et al. 2015; Robertson et al. 2015) and the microwave background data (Planck Collaboration et al. 2016). Thus, at least in CDM, the small observed number of faint galaxies sets a lower limit to the strength of feedback, while the requirement that the Universe be ionised early enough sets an upper limit. Hou et al. (2016) found that the simple models of SNfb usually assumed in semi-analytic models of galaxy formation do not satisfy both these requirements, because the default prescriptions are calibrated using $z = 0$ data and lack flexibility. They proposed instead a more complicated model in which the strength of SNfb evolves in redshift, as suggested by the SNfb model of Lagos et al. (2013) (see Sect. 4.2.2 below).

Since in WDM the number of small haloes is naturally suppressed, for a model to be viable, SNfb must be weak enough so that there are enough ionising photons at high redshift, as well as leading to the production of a sufficient number of satellite galaxies to account for observations in the Local Group.

4.2.2 Supernova Feedback in GALFORM

The observational data normally used to constrain and test semi-analytic models includes galaxies with stellar mass, $M_\star \gtrsim 10^8 M_\odot$. When attempting to extend the Lacey et al. (2016) model (which was used in Chap. 3) to lower mass galaxies, Hou et al. (2016) found that the original prescription for SNfb had to be modified as discussed in Sect. 4.2.1. In the original prescription, the mass loading factor, β, defined as the ratio of the mass ejection rate to the star formation rate, is assumed to be a power law in the circular velocity, V_{circ}, of the galaxy. To match the observed satellite luminosity function and produce an acceptable metallicity-luminosity relation for Milky Way satellites, Hou15 required a mass loading factor given by a broken power law with a redshift dependence:

$$
\beta = \begin{cases} (V_{\mathrm{circ}}/V_{\mathrm{SN}})^{-\gamma_{\mathrm{SN}}} & V_{\mathrm{circ}} \geq V_{\mathrm{thresh}} \\ (V_{\mathrm{circ}}/V'_{\mathrm{SN}})^{-\gamma'_{\mathrm{SN}}} & V_{\mathrm{circ}} < V_{\mathrm{thresh}}, \end{cases} \tag{4.2.4}
$$

where V'_{SN} is chosen such that the two power laws in Eq. 4.2.4 join at $V_{\mathrm{circ}} = V_{\mathrm{thresh}}$, $\gamma_{\mathrm{SN}} = 3.2$, $\gamma'_{\mathrm{SN}} = 1.0$, $V_{\mathrm{thresh}} = 50\,\mathrm{km s}^{-1}$ and:

$$
V_{\mathrm{SN}} = \begin{cases} 180 & z > 8 \\ -35z + 460 & 4 \leq z \leq 8 \,. \\ 320 & z < 4 \end{cases} \tag{4.2.5}
$$

This redshift dependence is chosen to capture the overall behaviour of Lagos et al. (2013) supernova feedback model. In the Hou et al. (2016) model, the feedback strength is assumed to be the same as in Lacey et al. (2016) at $z < 4$, but is weaker at higher redshifts and in galaxies with $V_{\mathrm{circ}} < V_{\mathrm{thresh}} = 50\,\mathrm{km s}^{-1}$. We will refer to this feedback scheme as the 'EvoFb' (evolving feedback) model.

The values of γ_{SN} and V_{thresh} in this model were calibrated for CDM and need to be recalibrated for the sterile neutrino models that we are considering. We find that the values $\gamma_{\mathrm{SN}} = 2.6$ for $L_6 = 700$, $\gamma_{\mathrm{SN}} = 2.8$ for $L_6 = (8, 12)$ and $V_{\mathrm{thresh}} = 30\,\mathrm{km s}^{-1}$

for all three values of L_6 provide the best-fit to the local b_J and K-band luminosity functions, the primary observables used to calibrate GALFORM.

4.2.3 Halo Merger Trees with Sterile Neutrinos

We generate merger trees using the extension of the Cole et al. (2000) Monte Carlo technique (based on the extended Press–Schechter (EPS) theory) described in Parkinson et al. (2008). In models in which the linear power spectrum, $P(k)$, has a cut-off, as in our sterile neutrino models, a small correction is required to the EPS formalism: to obtain the correct variance of the density field, $\sigma(M_{\mathrm{halo}})$, $P(k)$ needs to be convolved with a sharp k-space filter rather than with the real-space top-hat filter used for CDM (Benson et al. 2013). This choice results in good agreement with the conditional halo mass function obtained in N-body simulations (see, for example, Fig. 6 in Lovell et al. 2016).

Using our Monte Carlo technique rather than N-body simulations to generate merger trees has the advantage that different sterile neutrino models can be studied at minimum computational expense while avoiding the complication of spurious fragmentation in filaments that occurs in N-body simulations with a resolved cut-off in $P(k)$ (e.g. Wang and White 2007; Lovell et al. 2014).

4.3 Results

In this section, we present the main results of our models, consisting of predictions for field and satellite luminosity functions and the redshift of reionisation. We also investigate the sources that produce the ionising photons at high redshift.

4.3.1 Field Luminosity Functions

As discussed in Sect. 4.2.2, the parameters of the SNfb model in GALFORM were calibrated so as to obtain a good match to the present-day field galaxy luminosity functions. The b_J and K-band luminosity function in CDM and the $L_6 = (8, 12, 700)$ 7 keV sterile neutrino models are shown in Fig. 4.3. In both cases we have made use of the EvoFb feedback scheme of Sect. 4.2.2. We also consider an extreme model for $L_6 = 700$, in which supernova feedback is turned off completely ('NoFb'; photoionisation still occurs), thus maximising the amount of gas that is converted into stars.

In Fig. 4.3 we see that with the EvoFb scheme the observed luminosity functions are well reproduced in CDM and all our sterile neutrino models. This should come as no surprise since the EvoFb model parameters were tuned to match these particular

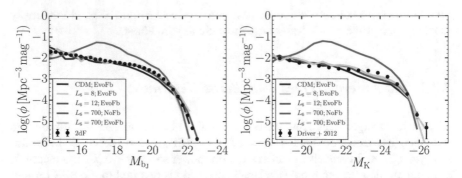

Fig. 4.3 The $z = 0$ field galaxy luminosity functions in the b_J-band (left panel) and the K-band (right panel) for the four dark matter models considered in this work: CDM and 7 keV sterile neutrino models with $L_6 = (8, 12, 700)$. The evolving feedback (EvoFb) model is used in GALFORM. For the $L_6 = 700$ case, we also show an extreme model in which the feedback has been completely turned off ('NoFb'). The black points are observational estimates (Norberg et al. 2002; Driver et al. 2012). Figure reproduced from Bose et al. (2016b)

data. The $L_6 = 700$ model, while inconsistent with the 3.5 keV line (see Table 4.1), is interesting because it has the most extreme power spectrum cut-off for a 7 keV sterile neutrino that produces the correct dark matter abundance. The maximum star formation efficiency in any model is obtained by turning off SNfb altogether. If in this limiting scenario the $L_6 = 700$ model produces too few faint galaxies to match the field luminosity function, this extreme model would be strongly ruled out. As Fig. 4.3 shows, the resultant luminosity function (shown in green) in fact overproduces faint galaxies.

4.3.2 Redshift of Reionisation

Since the onset of halo formation occurs later in sterile neutrino models compared to CDM (e.g. Bose et al. 2016a), star formation in dwarf galaxies is delayed (e.g. Colín et al. 2015, Governato et al. 2015, Bozek et al. 2018). Since, in addition, there are no haloes below a cut-off mass, it is unclear that enough sources of ionising photons will have formed to ionise hydrogen early enough to be consistent with the *Planck* limits on the redshift of reionisation (Planck Collaboration et al. 2016).

To answer this question we use GALFORM to calculate the ratio of the comoving number density of ionising photons produced, n_γ, to that of hydrogen nuclei, n_H as:

$$\mathscr{R}(z) = \frac{n_\gamma}{n_H} = \frac{\int_z^\infty \varepsilon(z')\,dz'}{n_H}, \tag{4.3.6}$$

where $\varepsilon(z')$ is the comoving number density of Lyman continuum photons produced per unit redshift. The Universe is deemed to be fully ionised at redshift $z_{\text{reion}}^{\text{full}}$ when the ratio in Eq. 4.3.6 reaches the value:

$$\mathcal{R}(z)|_{\text{full}} = \frac{1 + N_{\text{rec}}}{f_{\text{esc}}} = 6.25. \tag{4.3.7}$$

Here N_{rec} is the number of recombinations per hydrogen atom and f_{esc} is the fraction of ionising photons that are able to escape a galaxy into the IGM. Raičević et al. (2011) advocate a value of $N_{\text{rec}} = 1$ based on the hydrodynamical simulations of Iliev et al. (2006), Trac and Cen (2007). Finlator et al. (2012) suggest that photoheating would smooth the diffuse IGM and reduce the clumping factor by a factor of three compared with the value derived by Iliev et al. (2006). In this work, we will adopt a value $N_{\text{rec}} = 0.25$ (as in Hou15), but we have checked that our conclusions are insensitive to the exact value of this parameter. Furthermore, for the escape fraction, we assume $f_{\text{esc}} = 0.2$, which is consistent with the value used by Raičević et al. (2011). Sharma et al. (2016) present observational and theoretical evidence in support of this choice of f_{esc} (see also Khaire et al. 2016).

The microwave background data measure the optical depth to the time when the Universe (re)combined. This is usually converted into an equivalent 'redshift of reionisation' assuming a model of non-instantaneous reionisation. The value quoted in Planck Collaboration et al. (2016) corresponds to $z_{\text{reion}}^{\text{half}}$, the redshift at which the Universe is *half* ionised. With our assumptions this corresponds to:

$$\mathcal{R}(z)|_{\text{half}} = 3.125. \tag{4.3.8}$$

Reionisation suppresses galaxy formation in low mass haloes through an effect known as photoionisation feedback. In GALFORM, this is modelled using the approximation described in Benson et al. (2003): for haloes with virial velocity $V_{\text{vir}} < V_{\text{crit}}$, no gas cooling takes place for $z < z_{\text{crit}}$. As in Hou15, we adopt $z_{\text{crit}} = z_{\text{reion}}^{\text{full}}$ and $V_{\text{crit}} = 30\,\text{km s}^{-1}$ (Okamoto et al. 2008).

In the standard (Lacey et al. 2016) prescription, SNfb is modelled as a power law in the circular velocity of the galaxy without any dependence on redshift. Hou15 found that this model predicts $z_{\text{reion}}^{\text{half}} = 6.1$ for CDM, in conflict with the bounds by Planck Collaboration et al. (2016): $z_{\text{reion}}^{\text{half}} = 8.8^{+1.7}_{-1.4}$. We expect that sterile neutrino models, in which the formation of galaxies is both suppressed and delayed, would be in even greater conflict with the *Planck* observations. For this reason, in what follows we only consider the predictions of the evolving feedback (EvoFb) model of Hou15 (Sect. 4.2.2) which, at least for CDM, predicts an acceptable value for $z_{\text{reion}}^{\text{half}}$.

Figure 4.4 shows the evolution of $\mathcal{R}(z)$ with redshift for CDM and sterile neutrino models with $L_6 = (8, 12, 700)$ according to GALFORM with EvoFb feedback. In each panel, the intersection of the colour dashed lines marks $z_{\text{reion}}^{\text{half}}$, where $n_\gamma / n_{\text{H}} = 3.125$. The dashed grey line and shaded grey region mark the median and 68% confidence intervals from Planck Collaboration et al. (2016): $z_{\text{reion}}^{\text{half}} = 8.8^{+1.7}_{-1.4}$. In the bottom left of each panel, we give $z_{\text{reion}}^{\text{half}}$ and $z_{\text{reion}}^{\text{full}}$ predicted for each model.

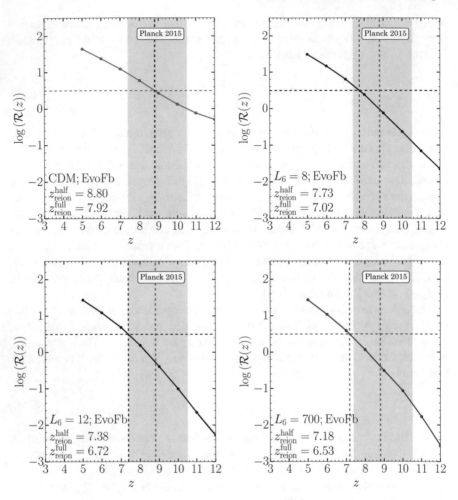

Fig. 4.4 The ratio of the total number of ionising photons produced up to redshift z as a fraction of the total comoving number density of hydrogen nuclei (solid lines in each panel). In each panel, we show the predictions for the different dark matter models under the EvoFb scheme. The intersection of the coloured dashed lines marks the redshift at which the Universe is 50% ionised; the redshifts for 50% ($z_{\mathrm{reion}}^{\mathrm{half}}$) and 100% reionisation ($z_{\mathrm{reion}}^{\mathrm{full}}$) are listed in the bottom left of each panel. The dashed grey line and shaded grey region demarcate the observational constraints as obtained from the *Planck* satellite, $z_{\mathrm{reion}}^{\mathrm{half}} = 8.8_{-1.4}^{+1.7}$ (at 68% confidence). Figure reproduced from Bose et al. (2016b)

All three 7 keV sterile neutrino models have values of $z_{\mathrm{reion}}^{\mathrm{half}}$ that are broadly consistent with the *Planck* data. The $L_6 = (12, 700)$ models fall just outside the lower 68% confidence lower limit and the $L_6 = 8$ model just inside. This is a non-trivial result given the paucity of early structure in these models compared to CDM. Unsurprisingly, $z_{\mathrm{reion}}^{\mathrm{half}}$ is higher in CDM.[2] Figure 4.4 already hints at the reason why

---

[2]We note that our results in this section contradict those by Rudakovskyi and Iakubovskyi (2016), who find that in the 7 keV $L_6 = 10$ model the Universe is reionised *earlier* than in CDM. This is

the sterile neutrino models are able to ionise the Universe early enough. Comparing, for example, the $L_6 = 700$ model (bottom right panel) to CDM (top left panel), it is clear that the evolution of $\log(\mathcal{R}(z))$ is steeper in the former, that is more UV photons are produced per unit redshift in the $L_6 = 700$ case, even though the *total* number of photons at that redshift is larger in CDM. For $L_6 = 8$, the most 'CDM-like' sterile neutrino model, the gradient of $\log(\mathcal{R}(z))$ is shallower. We will return to this feature shortly.

4.3.3 The Galaxies Responsible for Reionisation

We have seen that in spite of the delayed onset of galaxy formation, even the most extreme 7 keV sterile neutrino model is able to ionise the Universe early enough to be consistent with the constraints from *Planck*. To explore why this is so, we show in Fig. 4.5 several properties of the sources that contribute the bulk of the ionising photons at each redshift. Each column in the figure corresponds to a different dark matter model, while each row corresponds to a different property of the ionising sources: total stellar mass ($M_\star$, first row), halo mass (M_{halo}, second row) and galaxy circular velocity (V_{circ}, third row). The black vertical dashed lines mark $z_{\text{reion}}^{\text{full}}$, which is given in the top row in each case.

In CDM, the median stellar mass (i.e. the mass below which galaxies produce 50% of the ionising emissivity) at $z = z_{\text{reion}}^{\text{full}}$ is $\sim 10^8\,M_\odot$, whereas in the three sterile neutrino models the median mass is close to $\sim 10^9\,M_\odot$. The larger scatter in $M_\star$ and M_{halo} for CDM is due to the wide range of mass of the galaxies that contribute to the ionising photon budget. For example, at $z = 10$, galaxies with mass in the range $10^4\,M_\odot < M_\star < 10^9\,M_\odot$ contribute 90% of the ionising photons, whereas in the $L_6 = (12, 700)$ models, 90% of the photons are produced by galaxies with mass in the range $10^6\,M_\odot < M_\star < 10^9\,M_\odot$ since very few galaxies with $M_\star < 10^6\,M_\odot$ form in these models. The result is that the primary sources of ionising photons at high redshift in sterile neutrino are on average *more massive* than in CDM.

The build-up of the galaxy population in our models is illustrated in Fig. 4.6 which shows the rest frame far-UV (1500 Å) luminosity functions at $z = 7, 8, 9, 10$ in CDM and the $L_6 = (8, 12, 700)$ models. As noted in Hou15, in CDM the EvoFb feedback model predicts luminosity functions that are in good agreement with the data at all redshifts. EvoFb underpredicts the abundance of the brightest galaxies ($M_{\text{AB}}(1500\text{Å}) < -21$) for all dark matter models compared to the observations. For these galaxies, however, the data include many upper limits. Furthermore, these rare luminous galaxies are not the dominant sources of ionising photons (c.f. Fig. 4.5), so we do not expect the underprediction from the Hou15 model to impact our conclu-

---

ascribed to the lack of 'mini'-haloes in the sterile neutrino cosmology, which reduces the average number of recombinations per hydrogen atom. In our analysis this amounts to a reduction in the value of N_{rec} in Eq. 4.3.7. However, we have checked that even reducing the value of N_{rec} by a factor of 10 does not affect our results significantly.

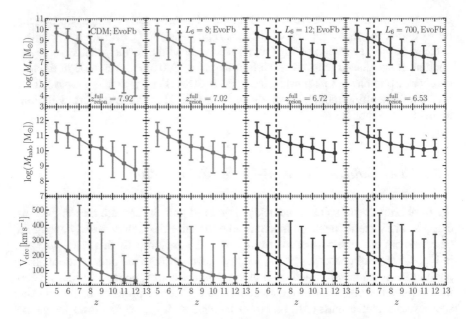

Fig. 4.5 Properties of the sources that produce ionising photons as a function of redshift for CDM and 7 keV sterile neutrino models with $L_6 = (8, 12, 700)$. The properties shown are stellar mass, $M_\star$ (top row), halo mass M_{halo} (middle row) and circular velocity (V_{circ}). The median (solid lines), 5th and 95th percentiles (error bars) are determined by weighting the contribution of each galaxy to the total ionising emissivity at that redshift. The black vertical dashed line in each case marks the redshift at which the Universe is half ionised. Figure reproduced from Bose et al. (2016b)

sions significantly for the redshift of reionisation in this work. For $L_6 = (12, 700)$, the models also underpredict the abundance of galaxies fainter than $M_{\text{AB}}(1500\text{Å}) \sim -20$ galaxies at $z = 9$ and $z = 10$. Reducing the strength of SNfb at $z > 8$ slightly can bring these models into agreement with the data without spoiling the agreement at $z = 0$.

An interesting feature of Fig. 4.6 is that while the $L_6 = (8, 12, 700)$ sterile neutrino models produce fewer galaxies fainter than $M_{\text{AB}}(1500\text{Å}) \sim -20$ at $z = 10$, all three models catch up with CDM by $z = 7$, roughly the time by which 50% hydrogen reionisation has occurred. The *build-up of the high redshift galaxies therefore proceeds more rapidly* in the sterile neutrino cosmologies than in CDM. This is consistent with the behaviour of the rate of ionising photon production seen in Sect. 4.3.2, where the slope of $\log(n_\gamma/n_{\text{H}})$ was shown to be steeper for sterile neutrino models compared to CDM.

The reason for the differing rates of galaxy formation at high redshift in the different models can be understood as follows. Due to the lack of progenitors below the cut-off mass scale, WDM haloes build up via roughly equal-mass mergers of intermediate mass haloes. Near the free streaming scale, the growth rate of haloes is therefore more rapid in WDM than in CDM (see, e.g. Ludlow et al. 2016). This

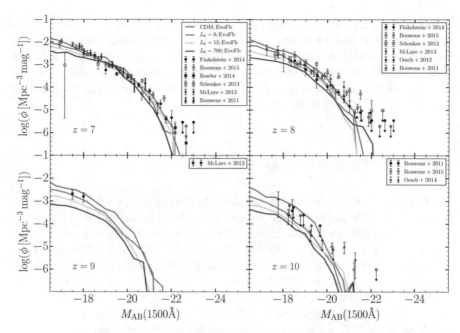

Fig. 4.6 Evolution of the rest frame far-UV galaxy luminosity functions from $z = 7 - 10$ in our models. The predictions of GALFORM for CDM and the $L_6 = (8, 12, 700)$ 7 keV sterile neutrino models are shown with solid colour lines as indicated in the legend. The symbols with errorbars are observational measurements (Bouwens et al. 2011a, b; Oesch et al. 2012; Schenker et al. 2013; McLure et al. 2013; Finkelstein et al. 2015; Bowler et al. 2014; Oesch et al. 2014; Bouwens et al. 2015). Figure reproduced from Bose et al. (2016b)

Table 4.2 The redshift interval during which the universe goes from 50% to 100% ionised. The values shown here are predictions using the EvoFb model

Model	$\Delta z_{1/2 \to 1}$
CDM	1.02
$L_6 = 8$	0.82
$L_6 = 12$	0.75
$L_6 = 700$	0.71

is why soon after the formation of the first galaxies the rate of galaxy formation in sterile neutrino models 'catches up' with the corresponding rate in CDM. This rapid early evolution, reflected for example in the UV luminosity function, is a generic prediction of WDM, independently of the details of the galaxy formation model.

The result of this rapid early phase of galaxy formation in WDM is that the actual *duration* of the Epoch of Reionisation (EoR) is *shorter* than in CDM. This is summarised in Table 4.2, where we tabulate the redshift interval during which the universe goes from 50 to 100% ionised. The values in this table assume the EvoFb model of feedback. While the magnitude of these values could be sensitive to the galaxy formation model assumed, the general trend – that warmer models exhibit a shorter EoR – should be independent of the choice of feedback model. Measurement of the kinetic Sunyaev–Zel'dovich effect or future 21 cm observations of the EoR that aim to constrain how fast reionisation proceeded could therefore place constraints on the nature of the dark matter itself.

4.3.4 Satellites of the Milky Way

The Milky Way satellite luminosity function has been used to set limits on the warm dark matter particle mass: if the power spectrum cut-off occurs on too large a scale, too few haloes form to account for the observed number of satellites (Macciò and Fontanot 2010; Polisensky and Ricotti 2011; Lovell et al. 2012; Nierenberg et al. 2013; Kennedy et al. 2014). These studies considered non-resonantly produced thermal relics (but see Schneider 2016). Lovell et al. (2016) considered sterile neutrino models, similar to ours, with different particle masses and values of L_6 and an earlier version of GALFORM (Gonzalez-Perez et al. 2014). There are degeneracies between the shape of the WDM power spectrum and some of the parameters of the galaxy formation model, particularly, of course, the strength of SNfb (see Kennedy et al. 2014 for a discussion). These degeneracies are mitigated in our case by considering a variety of observational constraints involving a range of halo masses and redshifts.

We have allowed the strength of SNfb to vary with redshift, by assuming that SNfb is weaker at high redshift. In Sect. 4.3.2, we found that this modification to the feedback scheme in GALFORM allows CDM and the $L_6 = (8, 12, 700)$ sterile neutrino models to reionise the Universe early enough to be consistent with the *Planck* limits on the redshift of reionisation. It is not clear, however, what the effect of reducing the strength of feedback will be on observables at lower redshifts. In particular, we expect the predicted luminosity function of satellites in the Milky Way to be particularly sensitive to this modification.

To predict the satellite luminosity functions around galaxies similar to the Milky Way we generate 100 Monte Carlo merger trees in 5 equally spaced bins of final halo masses in the range $5 \times 10^{11}\, M_\odot \leq M_{\rm halo}^{\rm host} \leq 2 \times 10^{12}\, M_\odot$. The cumulative V-band satellite luminosity functions at $z = 0$ are shown in Fig. 4.7 for our various dark matter models with the EvoFb feedback scheme. Before we attempt to compare these predictions with observations we note that the two different observational datasets plotted in the figure disagree with one another at the bright end of the luminosity function ($M_V \leq -8$), which is the regime of the 11 "classical" satellites. There are two reasons for this difference: firstly, McConnachie (2012), whose measurements are included in the bright end of the 'Combined data' sample includes Canis Major ($M_V = -14.4$), whereas this galaxy is excluded by Tollerud et al. (2008). Secondly, Tollerud et al. (2008) adopt $M_V = -9.8$ for Sculptor, compared to McConnachie's value of $M_V = -11.1$. At the faint end the differences in the satellite luminosity function arise from differing assumptions for the radial distributions of the satellites.

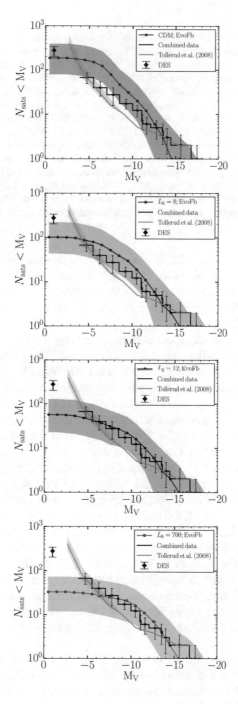

◀ **Fig. 4.7** Cumulative V-band Milky Way satellite luminosity functions at $z = 0$ for our four dark matter models with EvoFb supernova feedback. In each case, we have used 100 Monte Carlo merger trees for haloes of final mass in the range $5 \times 10^{11} - 2 \times 10^{12} M_{\odot}$. The smooth solid line indicates the median and the coloured shaded region the 5th and 95th percentiles over all realisations. The black histogram labelled 'Combined data' shows the observed Milky Way satellite luminosity function obtained by combining two datasets: for $M_V \geq -11$ the data are taken from Koposov et al. (2008), which includes corrections for incompleteness in the SDSS DR5 catalogue; for $M_V < -11$, the data are taken from McConnachie (2012). The solid grey line shows the estimated observed satellite luminosity function from Tollerud et al. (2008) with the grey shaded region showing the 98% spread over 18,576 mock surveys of the Milky Way halo in the Via Lactea simulation (Diemand et al. 2007). The black diamond marks an extension of the observed satellite luminosity function adding the new ultra-faint dwarf satellites discovered by DES down to $M_V \leq -1$ (Jethwa et al. 2016). The partial sky coverage of the survey is taken into account. All error bars are Poisson errors, including volume corrections where appropriate. Figure reproduced from Bose et al. (2016b)

In particular, Koposov et al. (2008) assume that the satellite distribution follows the NFW profile (Navarro et al. 1996, 1997) of the host halo, whereas Tollerud et al. (2008) assume the subhalo radial distribution measured in the Via Lactea simulations (Diemand et al. 2007). The radial distribution of subhaloes is similar in CDM and WDM (Bose et al. 2017).

Figure 4.7 shows that all of our models, including the most extreme $L_6 = 700$ case, are consistent with the data down to $M_V \sim -5$. For CDM the EvoFb model slightly overpredicts the number of the faintest satellites ($M_V > -8$), but here the data could be incomplete. However, since the number of satellites scales with the host halo mass (Wang et al. 2012b; Cautun et al. 2014), our sterile neutrino models would be increasingly in conflict with the observed luminosity functions for $M_{\mathrm{halo}}^{\mathrm{host}} \lesssim 10^{12} M_{\odot}$. For example, if $M_{\mathrm{halo}}^{\mathrm{host}} \leq 7 \times 10^{11} M_{\odot}$, both the $L_6 = 700$ and $L_6 = 12$ EvoFb models would be ruled out because they fail to form enough faint satellites with $M_V > -10$ even after accounting for the large scatter. Only CDM and our $L_6 = 8$ sterile neutrino models would remain consistent with the Koposov et al. (2008), McConnachie (2012) ('Combined data') observations in this case.

The Dark Energy Survey (DES) recently reported the discovery of new ultra-faint dwarf galaxies (Bechtol et al. 2015; Koposov et al. 2015; Drlica-Wagner et al. 2015; Jethwa et al. 2016). We can consider their contribution to the observed luminosity function following the analysis by Jethwa et al. (2016) who find that 12 of the 14 satellites have >50% probability of having been brought in as satellites of the LMC itself (at 95% confidence). Extrapolating from the detected population Jethwa et al. (2016) conclude that the Milky Way should have ~180 satellites within 300 kpc and 70^{+30}_{-40} Magellanic satellites in the magnitude range $-7 < M_V < -1$ (at 68% confidence).

The extrapolated contribution of the DES satellites (a total of 250 satellites) is represented by the black diamond in Fig. 4.7. CDM is consistent with this number particularly for the larger assumed values of the mass of the Milky Way halo. On the other hand, the 'coldest' 7 keV sterile neutrino, namely $L_6 = 8$, is only marginally

Fig. 4.8 Same as Fig. 4.7 for the $L_6 = 8$ model, but in an extreme scenario where feedback has been turned off completely. Figure reproduced from Bose et al. (2016b)

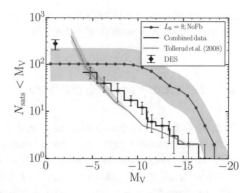

consistent with the extrapolation, while the $L_6 = 12$ and $L_6 = 700$ models are in significant disagreement with the extrapolated number count. The predicted number of faint dwarfs produced by any of these models is, of course, sensitive to the details of the SNfb but in the following section we consider a limiting case.

4.3.5 Model Independent Constraints on Dark Matter

As mentioned in Sect. 4.3.4 our analysis suffers from a degeneracy between the shape of the initial power spectrum and the strength of SNfb. A model independent constraint, however, can be derived by assuming that there is no SNfb at all. In this case, every subhalo in which gas can cool hosts a satellite, thus maximising the size of the population. In Fig. 4.8 we show the predicted Milky Way satellite luminosity function in the case of zero feedback ('NoFb'). The total number of satellites is determined entirely by reionisation i.e., by the amount of gas cooling in haloes prior to the onset of reionisation.

In Fig. 4.8 we have assumed $z_{\text{reion}}^{\text{full}} = 7.02$, as predicted by the EvoFb scheme for the $L_6 = 8$ model. This produces, on average, $\sim$100 satellites with $M_V \leq -1$. A fully self-consistent treatment of reionisation for the NoFb model would result in $z_{\text{reion}}^{\text{full}} > 7.02$, in which case the number of satellites produced would be even less than 100. The maximum number of satellite galaxies produced in Fig. 4.8 is converged with respect to the halo mass resolution. The figure shows that the extreme NoFb model L_6 is only marginally consistent with the extrapolated DES data for the $L_6 = 8$ case. We recall that this value of the lepton asymmetry corresponds to the 'coldest' possible 7 keV sterile neutrino; ruling this out would rule out the entire family of 7 keV sterile neutrinos as the dark matter particles.

The exact location of the extrapolated DES data point in the cumulative luminosity function is subject to a number of caveats, such as the DES selection function, detection efficiency and assumptions about isotropy. However, it is clear that the discovery of even more ultra-faint dwarf galaxies could potentially set very strong constraints on the nature of the dark matter.

4.4 Summary and Discussions

We have carried out a detailed investigation of the process of reionisation in models in which the dark matter particles are assumed to be sterile neutrinos. The free streaming of these particles leads to a sharp cut-off in the primordial matter power spectrum at the scale of dwarf galaxies (Fig. 4.1). On scales much larger than the cut-off, structure formation proceeds almost identically to CDM. Near and below the cut-off, sterile neutrinos behave like warm dark matter (WDM): the abundance of haloes (and therefore of the galaxies they host) is suppressed and their formation times are delayed relative to CDM. The sterile neutrino models we consider are motivated by observations of an X-ray excess at 3.5 keV in the stacked spectrum of galaxy clusters (Bulbul et al. 2014b) and in the spectra of M31 and the Perseus cluster (Boyarsky et al. 2014). This excess could be explained by the decay of a sterile neutrino with a rest mass of 7 keV.

In addition to their rest mass, sterile neutrinos are characterised by two additional parameters: the lepton asymmetry, L_6, and the mixing angle. Keeping the mass of the sterile neutrino fixed at 7 keV, we consider three values of L_6: 8, 12, 700. Based on their cut-off scales, the $L_6 = 8$ and $L_6 = 12$ models respectively correspond to the 'coldest' and 'warmest' 7 keV sterile neutrinos that are also consistent with the Bulbul et al. (2014b), Boyarsky et al. (2014) observations. The most extreme model we consider, $L_6 = 700$, also decays at 3.5 keV but the mixing angle is unable to produce a decay flux compatible with the 3.5 keV X-ray observations (see Table 4.1 for a summary).

To calculate the number of ionising photons produced in CDM and in the sterile neutrino models, we make use of the Durham semi-analytic model of galaxy formation, GALFORM using the supernova feedback prescription of Hou et al. (2016). In this model, the parameters controlling the strength and evolution of supernova feedback are calibrated for CDM by the epoch of reionisation as measured by *Planck*, and tested against data for the luminosity function and stellar mass-metallicity relation of Milky Way satellites (Sect. 4.2.2). We adopt similar values of the model parameters for our sterile neutrino models. Our main conclusions are:

(i) Although reionisation occurs slightly later in the sterile neutrino models than in CDM, the epoch of reionisation in all cases is consistent with the bounds from *Planck* (Sect. 4.3.2, Fig. 4.4). For the $L_6 = (12, 700)$ models, the redshifts at which the Universe is 50% ionised are just below the 68% confidence interval from *Planck*. Reionisation in the $L_6 = 8$ model occurs well within the *Planck* limits.

(ii) The galaxies that account for the bulk of the ionising photon budget are more massive in sterile neutrino models than in CDM (Sect. 4.3.3, Fig. 4.5). By the time reionisation is complete, 50% of the photoionising budget is produced by

$M_\star \lesssim 10^8 \, M_\odot$ galaxies in CDM; the median stellar mass is $M_\star \sim 10^9 \, M_\odot$ for the sterile neutrino models.

(iii) From the evolution of the far-UV luminosity function, we infer that the galaxy population at high redshift ($z > 7$) builds up more rapidly in the sterile neutrino models than in CDM (Sect. 4.3.3, Fig. 4.6). This is particularly pronounced in the case of the most extreme model, $L_6 = 700$, which produces far fewer galaxies than CDM at $z = 10$ but 'catches up' with the CDM UV luminosity function by $z = 7$. This is directly related to the more rapid mass accretion of haloes near the free streaming scale in WDM than in CDM. The qualitative difference in the growth of high redshift galaxies between CDM and WDM models does not depend on the details of the galaxy formation model.

(iv) CDM, as well as the three sterile neutrino models we have considered, are in good agreement with the present-day luminosity function of the "classical" and SDSS Milky Way satellite galaxies (Sect. 4.3.4, Fig. 4.7). For larger values of the mass of the Milky Way halo ($M_{\rm halo}^{\rm host} > 1 \times 10^{12} M_\odot$), even the $L_6 = 700$ model is consistent with the observations of Koposov et al. (2008), McConnachie (2012). On the other hand, if $M_{\rm halo}^{\rm host} \leq 7 \times 10^{11} M_\odot$, both the $L_6 = 700$ and $L_6 = 12$ models can be ruled out.

(v) Extrapolating to the whole sky the abundance of ultra-faint Milky Way dwarf satellite galaxies recently detected by DES extends that satellite luminosity function to very faint magnitudes. With this extrapolation, the sheer number of satellites places strong constraints on the sterile neutrino models which produce only a limited number of substructures. CDM is consistent with this extrapolation, but the 'coldest' 7 keV sterile neutrino (the $L_6 = 8$ model) is only marginally in agreement even when feedback is turned off completely, a limiting model in which the satellite population is maximised. Ruling out the $L_6 = 8$ model, the coolest of the 7keV sterile neutrino family, would rule out this entire class as candidates for the dark matter. However, extrapolating the DES counts to infer the total number of satellites is still subject to a number of assumptions and uncertainties.

4.5 New Developments Since Submission of This Thesis in April 2017

The high redshift universe is clearly an interesting regime for exploring the nature of dark matter. As we have seen in the present chapter (as well as Chaps. 2 and 3), the early free streaming of WDM particles results in a delayed start to the formation of the first stars and galaxies relative to CDM. While this seemingly has little impact on when the Universe was reionised ($z \sim 8$–10, see Sect. 4.3.2), it is interesting to consider what happens at even high redshift.

Very recently, the Experiment to Detect the Global Epoch of Reionization Signatures (EDGES, Bowman et al. 2018) detected a strong absorption feature centred at $\nu = 78 \pm 1$ MHz, possibly associated with the absorption of 21 cm CMB photons by neutral hydrogen at $z \sim 15-20$. This absorption may be caused UV light from the first generation of stars, which decouple neutral hydrogen from CMB photons through the so-called Wouthuysen-Field effect. The timing of this signal, therefore, poses an interesting problem for WDM scenarios: if the first sources of UV photons are not already present at these redshifts, the EDGES detection, if confirmed, would place very strong constraints on the lower limit to the WDM particle mass.

Schneider (2018) presents one such analysis for thermal WDM/sterile neutrino scenarios in light of the EDGES detection. Using analytical prescriptions for the abundance of haloes at these epochs, the analysis then folds in simple models for the star formation rate and flux of Lyman-α photons from these haloes. In doing so, Schneider (2018) finds that when combined with constraints from X-ray data, the EDGES detection, taken at face value, places very strong constraints on models where the dark matter is composed purely of warm particles, potentially ruling out the entire class of 7 keV sterile neutrinos. For 'mixed' dark matter scenarios (comprised of some combination of cold and warm/hot sub-components) the fraction of the non-cold component is constrained to 17% or less. In future work, it will be interesting to compare these predictions with hydrodynamical simulations that self-consistently couple the gravitational collapse of sterile neutrinos with star formation and radiative transfer.

One caveat to the analysis by Schneider (2018) is regarding the formation sites of the first stars in WDM, which in the present analysis has been assumed to be in haloes. In fact, hydrodynamical simulations performed by Gao and Theuns (2007) show that the first stars are in fact formed in *filaments* rather than haloes. Whereas in small-scale power in CDM result in the fragmentation of filaments into mini-haloes. The absence of small-scale power in WDM allows the gas in filaments to cool efficiently and get dense enough to form stars – long before the non-linear collapse of haloes has taken place. In fact, Gao et al. (2015) find that in WDM, star formation in filaments dominates that in haloes until $z \sim 6$. Including this effect could have a substantial impact on the constraint derived for WDM-like models using 21 cm cosmology. Confirmation of the EDGES detection will be of great value.

References

Abazajian KN (2014) Phys Rev Lett 112:161303. https://doi.org/10.1103/PhysRevLett.112.161303, http://adsabs.harvard.edu/abs/2014PhRvL.112p1303A

Avila-Reese V, Colín P, Valenzuela O, D'Onghia E, Firmani C (2001) ApJ 559:516. https://doi.org/10.1086/322411, http://adsabs.harvard.edu/abs/2001ApJ...559..516A

Baur J, Palanque-Delabrouille N, Yèche C, Magneville C, Viel M (2016) JCAP 8:012. https://doi.org/10.1088/1475-7516/2016/08/012, http://adsabs.harvard.edu/abs/2016JCAP...08..012B

Bechtol K et al (2015) ApJ 807:50. https://doi.org/10.1088/0004-637X/807/1/50, http://adsabs.harvard.edu/abs/2015ApJ...807...50B

Benson AJ, Frenk CS, Baugh CM, Cole S, Lacey CG (2003) MNRAS 343:679. https://doi.org/10.1046/j.1365-8711.2003.06709.x, http://adsabs.harvard.edu/abs/2003MNRAS.343..679B

Benson AJ et al (2013) MNRAS 428:1774. https://doi.org/10.1093/mnras/sts159, http://adsabs.harvard.edu/abs/2013MNRAS.428.1774B

Bode P, Ostriker JP, Turok N (2001) ApJ 556:93 https://doi.org/10.1086/321541, http://adsabs.harvard.edu/abs/2001ApJ...556...93B

Bose S, Hellwing WA, Frenk CS, Jenkins A, Lovell MR, Helly JC, Li B (2016a) MNRAS 455:318. https://doi.org/10.1093/mnras/stv2294, http://adsabs.harvard.edu/abs/2016MNRAS.455..318B

Bose S, Frenk CS, Hou J, Lacey CG, Lovell MR (2016b) MNRAS 463:3848. https://doi.org/10.1093/mnras/stw2288, http://adsabs.harvard.edu/abs/2016MNRAS.463.3848B

Bose S et al (2017) MNRAS 464:4520. https://doi.org/10.1093/mnras/stw2686, http://adsabs.harvard.edu/abs/2017MNRAS.464.4520B

Bouwens RJ et al (2011a) Nature 469:504. https://doi.org/10.1038/nature09717, http://adsabs.harvard.edu/abs/2011Natur.469..504B

Bouwens RJ et al (2011b) ApJ 737:90. https://doi.org/10.1088/0004-637X/737/2/90, http://adsabs.harvard.edu/abs/2011ApJ...737...90B

Bouwens RJ et al (2015) ApJ 803:34. https://doi.org/10.1088/0004-637X/803/1/34, http://adsabs.harvard.edu/abs/2015ApJ...803...34B

Bowler RAA et al (2014) MNRAS 440:2810. https://doi.org/10.1093/mnras/stu449, http://adsabs.harvard.edu/abs/2014MNRAS.440.2810B

Bowman JD, Rogers AEE, Monsalve RA, Mozdzen TJ, Mahesh N (2018) Nature 555:67. https://doi.org/10.1038/nature25792, http://adsabs.harvard.edu/abs/2018Natur.555...67B

Boyarsky A, Lesgourgues J, Ruchayskiy O, Viel M (2009a) JCAP 5:012. https://doi.org/10.1088/1475-7516/2009/05/012, http://adsabs.harvard.edu/abs/2009JCAP...05..012B

Boyarsky A, Lesgourgues J, Ruchayskiy O, Viel M (2009b) Phys Rev Lett 102:201304. https://doi.org/10.1103/PhysRevLett.102.201304, http://adsabs.harvard.edu/abs/2009PhRvL.102t1304B

Boyarsky A, Ruchayskiy O, Iakubovskyi D, Franse J (2014) Phys Rev Lett 113:251301. https://doi.org/10.1103/PhysRevLett.113.251301, http://adsabs.harvard.edu/abs/2014PhRvL.113y1301B

Bozek B, Boylan-Kolchin M, Horiuchi S, Garrison-Kimmel S, Abazajian K, Bullock JS (2016) MNRAS 459:1489. https://doi.org/10.1093/mnras/stw688, http://adsabs.harvard.edu/abs/2016MNRAS.459.1489B

Bozek B et al (2018). arXiv: 1803.05424

Bulbul E, Markevitch M, Foster A, Smith RK, Loewenstein M, Randall SW (2014b) ApJ 789:13. https://doi.org/10.1088/0004-637X/789/1/13, http://adsabs.harvard.edu/abs/2014ApJ...789...13B

Calura F, Menci N, Gallazzi A (2014) MNRAS 440:2066. https://doi.org/10.1093/mnras/stu339, http://adsabs.harvard.edu/abs/2014MNRAS.440.2066C

Cautun M, Hellwing WA, van de Weygaert R, Frenk CS, Jones BJT, Sawala T (2014) MNRAS 445:1820. https://doi.org/10.1093/mnras/stu1829, http://adsabs.harvard.edu/abs/2014MNRAS.445.1820C

Cole S, Aragon-Salamanca A, Frenk CS, Navarro JF, Zepf SE (1994) MNRAS 271:781. http://adsabs.harvard.edu/abs/1994MNRAS.271..781C

Cole S, Lacey CG, Baugh CM, Frenk CS (2000) MNRAS 319:168. https://doi.org/10.1046/j.1365-8711.2000.03879.x, http://adsabs.harvard.edu/abs/2000MNRAS.319..168C

Colín P, Avila-Reese V, Valenzuela O (2000) ApJ 542: 622. https://doi.org/10.1086/317057, http://adsabs.harvard.edu/abs/2000ApJ...542..622C

Colín P, Avila-Reese V, González-Samaniego A, Velázquez H (2015) ApJ 803:28. https://doi.org/10.1088/0004-637X/803/1/28, http://adsabs.harvard.edu/abs/2015ApJ...803...28C

Dayal P, Mesinger A, Pacucci F (2015) ApJ 806:67. https://doi.org/10.1088/0004-637X/806/1/67, http://adsabs.harvard.edu/abs/2015ApJ...806...67D

Dayal P, Choudhury TR, Bromm V, Pacucci F (2017) ApJ 836:16. https://doi.org/10.3847/1538-4357/836/1/16, http://adsabs.harvard.edu/abs/2017ApJ...836...16D

Diemand J, Kuhlen M, Madau P (2007) ApJ 657:262. https://doi.org/10.1086/510736, http://adsabs.harvard.edu/abs/2007ApJ...657..262D

Driver SP et al (2012) MNRAS 427:3244. https://doi.org/10.1111/j.1365-2966.2012.22036.x, http://adsabs.harvard.edu/abs/2012MNRAS.427.3244D

Drlica-Wagner A et al (2015) ApJ 813:109. https://doi.org/10.1088/0004-637X/813/2/109, http://
    adsabs.harvard.edu/abs/2015ApJ...813..109D
Efstathiou G (1992) MNRAS 256:43p. http://adsabs.harvard.edu/abs/1992MNRAS.256P..43E
Finkelstein SL et al (2015) ApJ 810:71. https://doi.org/10.1088/0004-637X/810/1/71, http://
    adsabs.harvard.edu/abs/2015ApJ...810...71F
Finlator K, Oh SP, Özel F, Davé R (2012) MNRAS 427:2464. https://doi.org/10.1111/j.1365-2966.
    2012.22114.x, http://adsabs.harvard.edu/abs/2012MNRAS.427.2464F
Gao L, Theuns T (2007) Science 317:1527. https://doi.org/10.1126/science.1146676, http://adsabs.
    harvard.edu/abs/2007Sci...317.1527G
Gao L, Theuns T, Springel V (2015) MNRAS 450:45. https://doi.org/10.1093/mnras/stv643, http://
    adsabs.harvard.edu/abs/2015MNRAS.450...45G
Garzilli A, Boyarsky A, Ruchayskiy O (2015). arXiv: 1510.07006
Ghiglieri J, Laine M (2015) J High Energy Phys 11:171. https://doi.org/10.1007/JHEP11(2015)171,
    http://adsabs.harvard.edu/abs/2015JHEP...11..171G
Gonzalez-Perez V, Lacey CG, Baugh CM, Lagos CDP, Helly J, Campbell DJR, Mitchell PD
    (2014) MNRAS 439:264. https://doi.org/10.1093/mnras/stt2410, http://adsabs.harvard.edu/abs/
    2014MNRAS.439..264G
Governato F et al (2015) MNRAS 448:792. https://doi.org/10.1093/mnras/stu2720, http://adsabs.
    harvard.edu/abs/2015MNRAS.448..792G
Hou J, Frenk CS, Lacey CG, Bose S (2016) MNRAS 463:1224. https://doi.org/10.1093/mnras/
    stw2033, http://adsabs.harvard.edu/abs/2016MNRAS.463.1224H
Iliev IT, Mellema G, Pen U-L, Merz H, Shapiro PR, Alvarez MA (2006) MNRAS
    369:1625. https://doi.org/10.1111/j.1365-2966.2006.10502.x, http://adsabs.harvard.edu/abs/
    2006MNRAS.369.1625I
Jenkins A, Frenk CS, White SDM, Colberg JM, Cole S, Evrard AE, Couchman HMP, Yoshida
    N (2001) MNRAS 321:372. https://doi.org/10.1046/j.1365-8711.2001.04029.x, http://adsabs.
    harvard.edu/abs/2001MNRAS.321..372J
Jethwa P, Erkal D, Belokurov V (2016) MNRAS 461:2212. https://doi.org/10.1093/mnras/stw1343,
    http://adsabs.harvard.edu/abs/2016MNRAS.461.2212J
Kauffmann G, White SDM, Guiderdoni B (1993) MNRAS 264:201. https://doi.org/10.1093/mnras/
    264.1.201, http://adsabs.harvard.edu/abs/1993MNRAS.264..201K
Kennedy R, Frenk C, Cole S, Benson A (2014) MNRAS 442:2487. https://doi.org/10.1093/mnras/
    stu719, http://adsabs.harvard.edu/abs/2014MNRAS.442.2487K
Khaire V, Srianand R, Choudhury TR, Gaikwad P (2016) MNRAS 457:4051. https://doi.org/10.
    1093/mnras/stw192, http://adsabs.harvard.edu/abs/2016MNRAS.457.4051K
Koposov S et al (2008) ApJ 686:279. https://doi.org/10.1086/589911, http://adsabs.harvard.edu/
    abs/2008ApJ...686..279K
Koposov SE, Belokurov V, Torrealba G, Evans NW (2015) ApJ 805:130. https://doi.org/10.1088/
    0004-637X/805/2/130, http://adsabs.harvard.edu/abs/2015ApJ...805..130K
Lacey CG et al (2016) MNRAS 462:3854. https://doi.org/10.1093/mnras/stw1888, http://adsabs.
    harvard.edu/abs/2016MNRAS.462.3854L
Lagos CdP, Lacey CG, Baugh CM (2013) MNRAS 436:1787. https://doi.org/10.1093/mnras/
    stt1696, http://adsabs.harvard.edu/abs/2013MNRAS.436.1787L
Laine M, Shaposhnikov M (2008) JCAP 6:31. https://doi.org/10.1088/1475-7516/2008/06/031,
    http://adsabs.harvard.edu/abs/2008JCAP...06..031L
Larson RB (1974) MNRAS 169:229. http://adsabs.harvard.edu/abs/1974MNRAS.169..229L
Lewis A, Challinor A, Lasenby A (2000) ApJ 538:473. https://doi.org/10.1086/309179, http://
    adsabs.harvard.edu/abs/2000ApJ...538..473L
Lovell MR et al (2012) MNRAS 420:2318. https://doi.org/10.1111/j.1365-2966.2011.20200.x,
    http://adsabs.harvard.edu/abs/2012MNRAS.420.2318L
Lovell MR, Frenk CS, Eke VR, Jenkins A, Gao L, Theuns T (2014) MNRAS 439:300. https://doi.
    org/10.1093/mnras/stt2431, http://adsabs.harvard.edu/abs/2014MNRAS.439..300L

Lovell MR et al (2016) MNRAS 461:60. https://doi.org/10.1093/mnras/stw1317, http://adsabs.harvard.edu/abs/2016MNRAS.461...60L

Ludlow AD, Bose S, Angulo RE, Wang L, Hellwing WA, Navarro JF, Cole S, Frenk CS (2016) MNRAS 460:1214. https://doi.org/10.1093/mnras/stw1046, http://adsabs.harvard.edu/abs/2016MNRAS.460.1214L

Macciò AV, Fontanot F (2010) MNRAS 404:L16. https://doi.org/10.1111/j.1745-3933.2010.00825.x, http://adsabs.harvard.edu/abs/2010MNRAS.404L..16M

Madau P, Haardt F (2015) ApJL 813:L8. https://doi.org/10.1088/2041-8205/813/1/L8, http://adsabs.harvard.edu/abs/2015ApJ...813L...8M

Maio U, Viel M (2015) MNRAS 446:2760. https://doi.org/10.1093/mnras/stu2304, http://adsabs.harvard.edu/abs/2015MNRAS.446.2760M

McConnachie AW (2012) AJ 144:4. https://doi.org/10.1088/0004-6256/144/1/4, http://adsabs.harvard.edu/abs/2012AJ....144....4M

McLure RJ et al (2013) MNRAS 432: 2696. https://doi.org/10.1093/mnras/stt627, http://adsabs.harvard.edu/abs/2013MNRAS.432.2696M

Mitra S, Choudhury TR, Ferrara A (2015) MNRAS 454:L76. https://doi.org/10.1093/mnrasl/slv134, http://adsabs.harvard.edu/abs/2015MNRAS.454L..76M

Navarro JF, Frenk CS, White SDM (1996) ApJ 462:563. https://doi.org/10.1086/177173, http://adsabs.harvard.edu/abs/1996ApJ...462..563N

Navarro JF, Frenk CS, White SDM (1997) ApJ 490:493. http://adsabs.harvard.edu/abs/1997ApJ...490..493N

Nierenberg AM, Treu T, Menci N, Lu Y, Wang W (2013) ApJ 772:146. https://doi.org/10.1088/0004-637X/772/2/146, http://adsabs.harvard.edu/abs/2013ApJ...772..146N

Norberg P et al (2002) MNRAS 336:907. https://doi.org/10.1046/j.1365-8711.2002.05831.x, http://adsabs.harvard.edu/abs/2002MNRAS.336..907N

Oesch PA et al (2012) ApJ 759:135. https://doi.org/10.1088/0004-637X/759/2/135, http://adsabs.harvard.edu/abs/2012ApJ...759..135O

Oesch PA et al (2014) ApJ 786: 108. https://doi.org/10.1088/0004-637X/786/2/108, http://adsabs.harvard.edu/abs/2014ApJ...786..108O

Okamoto T, Gao L, Theuns T (2008) MNRAS 390:920. https://doi.org/10.1111/j.1365-2966.2008.13830.x, http://adsabs.harvard.edu/abs/2008MNRAS.390..920O

Parkinson H, Cole S, Helly J (2008) MNRAS 383:557. https://doi.org/10.1111/j.1365-2966.2007.12517.x, http://adsabs.harvard.edu/abs/2008MNRAS.383..557P

Planck Collaboration et al (2016) A&A 594:A13. https://doi.org/10.1051/0004-6361/201525830, http://adsabs.harvard.edu/abs/2016A%26A...594A..13P

Polisensky E, Ricotti M (2011) Phys Rev D 83:043506. https://doi.org/10.1103/PhysRevD.83.043506, http://adsabs.harvard.edu/abs/2011PhRvD..83d3506P

Raičević M, Theuns T, Lacey C (2011) MNRAS 410:775. https://doi.org/10.1111/j.1365-2966.2010.17480.x, http://adsabs.harvard.edu/abs/2011MNRAS.410..775R

Robertson BE, Ellis RS, Furlanetto SR, Dunlop JS (2015) ApJL 802:L19. https://doi.org/10.1088/2041-8205/802/2/L19, http://adsabs.harvard.edu/abs/2015ApJ...802L..19R

Rudakovskyi A, Iakubovskyi D (2016) JCAP 6:017. https://doi.org/10.1088/1475-7516/2016/06/017, http://adsabs.harvard.edu/abs/2016JCAP...06..017R

Sawala T et al (2015) MNRAS 448:2941. https://doi.org/10.1093/mnras/stu2753, http://adsabs.harvard.edu/abs/2015MNRAS.448.2941S

Schenker MA et al (2013) ApJ 768:196. https://doi.org/10.1088/0004-637X/768/2/196, http://adsabs.harvard.edu/abs/2013ApJ...768..196S

Schneider A (2016) JCAP 4:059. https://doi.org/10.1088/1475-7516/2016/04/059, http://adsabs.harvard.edu/abs/2016JCAP...04..059S

Schneider A (2018). arXiv: 1805.00021

Schneider A, Smith RE, Macciò AV, Moore B (2012) MNRAS 424:684. https://doi.org/10.1111/j.1365-2966.2012.21252.x, http://adsabs.harvard.edu/abs/2012MNRAS.424..684S

Schultz C, Oñorbe J, Abazajian KN, Bullock JS (2014) MNRAS 442:1597. https://doi.org/10.1093/mnras/stu976, http://adsabs.harvard.edu/abs/2014MNRAS.442.1597S

Sharma M, Theuns T, Frenk C, Bower R, Crain R, Schaller M, Schaye J (2016) MNRAS. https://doi.org/10.1093/mnrasl/slw021, http://adsabs.harvard.edu/abs/2016MNRAS.tmpL...4S

Sheth RK, Tormen G (1999) MNRAS 308:119. https://doi.org/10.1046/j.1365-8711.1999.02692.x, http://adsabs.harvard.edu/abs/1999MNRAS.308..119S

Tinker J, Kravtsov AV, Klypin A, Abazajian K, Warren M, Yepes G, Gottlöber S, Holz DE (2008) ApJ 688:709. https://doi.org/10.1086/591439, http://adsabs.harvard.edu/abs/2008ApJ...688..709T

Tollerud EJ, Bullock JS, Strigari LE, Willman B (2008) ApJ 688:277. https://doi.org/10.1086/592102, http://adsabs.harvard.edu/abs/2008ApJ...688..277T

Trac H, Cen R (2007) ApJ 671:1. https://doi.org/10.1086/522566, http://adsabs.harvard.edu/abs/2007ApJ...671....1T

Viel M, Lesgourgues J, Haehnelt MG, Matarrese S, Riotto A (2005) Phys Rev D 71:063534. https://doi.org/10.1103/PhysRevD.71.063534, http://adsabs.harvard.edu/abs/2005PhRvD..71f3534V

Viel M, Becker GD, Bolton JS, Haehnelt MG (2013) Phys Rev D, 88:043502. https://doi.org/10.1103/PhysRevD.88.043502, http://adsabs.harvard.edu/abs/2013PhRvD..88d3502V

Wang J, White SDM (2007) MNRAS 380:93. https://doi.org/10.1111/j.1365-2966.2007.12053.x, http://adsabs.harvard.edu/abs/2007MNRAS.380...93W

Wang J, Frenk CS, Navarro JF, Gao L, Sawala T (2012b) MNRAS 424:2715. https://doi.org/10.1111/j.1365-2966.2012.21357.x, http://adsabs.harvard.edu/abs/2012MNRAS.424.2715W

White SDM, Frenk CS (1991) ApJ 379:52. https://doi.org/10.1086/170483, http://adsabs.harvard.edu/abs/1991ApJ...379...52W

Part II
Numerical Simulations of f(R) Gravity

Chapter 5
Testing the Quasi-static Approximation in f(R) Gravity Simulations

5.1 Introduction

[1]In recent years, theories of modified gravity have become a subject of great interest in alternative approaches to modelling the observed acceleration of the Universe (Riess et al. 1998; Perlmutter et al. 1999). Einstein's theory of General Relativity (GR) has been the underlying gravity theory in the standard cosmological model of ΛCDM, the dark energy (Λ) and (cold) dark matter (CDM) components of which remain unresolved challenges to cosmologists. Modified gravity seeks to answer this question by modifying the theory of gravity itself, most routinely with the addition of scalar, vector or tensorial modifications to the Einstein–Hilbert action that governs GR (see Clifton et al. 2012, for a comprehensive review). Of course, one cannot deny the undoubted success of GR in passing local and Solar System tests of gravity, and so it is necessary for any reasonable modified gravity theory to also do the same. One process by which a modified theory reduces to GR on small scales is known as *screening*, Khoury (2010) of which there are three main types: chameleon (Shaw 2007), Vainshtein (Dvali et al. 2000) and dilaton/symmetron screening (Hinterbichler and Khoury 2010; Brax et al. 2010), with different theories equipped with different screening mechanisms.

One of the most popular models of modified gravity is $f(R)$ gravity (Carroll et al. 2005). This theory is built around the addition of a scalar function of the Ricci curvature scalar to the Einstein–Hilbert action. The scalar field has a potential, which acts as an effective cosmological constant that accelerates the expansion of the Universe, and also generates a 'fifth force' between matter particles. While the fifth force enhances the standard Newtonian gravity in low-density regions, high-density regions, GR is recovered by means of the chameleon screening. This mechanism is a consequence of the high degree of non-linearity in the equations of motion that

[1]The content of this chapter is based on the article Bose et al. 'Testing the quasi-static approximation in $f(R)$ gravity simulations', Journal of Cosmology and Astroparticle Physics, Volume 2015, Issue 2, published 24 February 2015. Reproduced with permission. All rights reserved, https://doi.org/10.1088/1475-7516/2015/02/034.

© Springer International Publishing AG, part of Springer Nature 2018
S. Bose, *Beyond ΛCDM* , Springer Theses,
https://doi.org/10.1007/978-3-319-96761-5_5

govern this theory. Its presence makes standard perturbative approaches less useful, and calls for the need to perform N-body simulations at high-resolution to fully understand the cosmological behaviour of this model.

Numerical simulations for $f(R)$ gravity (and for most other modified gravity theories) have traditionally been performed in what is known as the "quasi-static limit", in which the time derivatives of the scalar field that generates the fifth force are considered small compared to its spatial derivatives, and can therefore be safely neglected (Li et al. 2012; Cai et al. 2014; Hellwing et al. 2013). An advantage of this approximation is that it considerably simplifies the challenge of numerically solving the non-linear equations. In GR simulations, this approximation has been tested as being valid, but while it is consistently made in the case of $f(R)$ simulations, its validity has not yet been tested rigorously, especially in the non-linear regime (we note that recently efforts to include non-static effects have been made in the case of symmetron screening, Llinares and Mota 2014a). Testing the validity of this approximation is imperative given how widely-studied the $f(R)$ model is.

The aim of our investigation here is to quantitatively estimate the effects of excluding the time derivatives in N-body simulations for $f(R)$ gravity. For this purpose, we have derived field equations in which time derivatives of the scalar field are consistently included, and implemented these equations in a modified version of the ECOSMOG code (Li et al. 2012). By running simulations at different resolutions, we then study how the clustering of matter is affected by the non-static effects. We find that in low-resolution simulations, the time derivatives do have an impact on the observables we study, but this diminishes when we re-simulate at higher resolution or shorter time steps. As a result, at least for the $f(R)$ models we have studied, the quasi-static approximation seems to be valid for the observables we are interested in.

This chapter is organised as follows: in Sect. 5.2, we introduce the Hu–Sawicki (2007) $f(R)$ model, and how chameleon screening is able to recover GR. Sections. 5.3 and 5.4 describe how we modify the ordinary evolution equations to account for time derivatives in the non-linear regime, and how these equations are then discretised for the purpose of solving them on a mesh. In Sect. 5.5, we present the results of our N-body simulations at different resolutions, while in Sect. 5.6, we discuss some numerical aspects that must be taken into account when interpreting the results of our work. Finally, in Sect. 5.7, we summarise our findings and their implications.

Throughout this chapter, Greek indices run over 0, 1, 2, 3 (the four space-time components) whereas Latin indices run over 1, 2, 3 (the three spatial components).

5.2 An Introduction to $f(R)$ Gravity

In this section, we will briefly discuss the main features of $f(R)$ gravity, first in general, and then with the more specific example of the Hu–Sawicki (2007) model, which is the one we will analyse in the rest of this chapter. We expect that our findings in this work are at least qualitatively applicable to other classes of $f(R)$ models as well.

5.2.1 $f(R)$ Gravity: An Overview

As with most modified gravity theories, the starting point is the Einstein–Hilbert action. The modification we make is to replace the cosmological constant Λ with a function of the Ricci scalar, R, as:

$$S = \int \mathrm{d}^4x\sqrt{-g}\left[\frac{1}{2}M_{\mathrm{Pl}}^2\,[R + f(R)] + \mathscr{L}_m\right], \tag{5.2.1}$$

where g is the determinant of the metric tensor $g_{\mu\nu}$, $M_{\mathrm{Pl}} = 1/\sqrt{8\pi G}$ is the reduced Planck mass, G is the Newtonian gravitational constant, and $\mathscr{L}_m$ the total matter (baryonic + dark matter) Lagrangian density. We assume that neutrinos are massless, and that at late times the contribution from photons and neutrinos is negligible. The distinction between different $f(R)$ models is in the specific choice for the function $f(R)$ itself.

By varying the action in Eq. 5.2.1 with respect to the metric $g_{\mu\nu}$, we obtain the modified Einstein field equations:

$$G_{\mu\nu} + f_R R_{\mu\nu} - \left[\frac{1}{2}f(R) - \Box f_R\right]g_{\mu\nu} - \nabla_\mu\nabla_\nu f_R$$
$$= 8\pi G T^{\mathrm{m}}_{\mu\nu}, \tag{5.2.2}$$

where $G_{\mu\nu} = R_{\mu\nu} - \frac{1}{2}g_{\mu\nu}R$ is the Einstein tensor, ∇_μ is the covariant derivative compatible with the metric $g_{\mu\nu}$, $\Box \equiv \nabla^\mu\nabla_\mu$, $T^{\mathrm{m}}_{\mu\nu}$ is the energy-momentum tensor for matter, and $f_R \equiv \frac{\mathrm{d}f(R)}{\mathrm{d}R}$ is the extra scalar degree of freedom of this model, known as the *scalaron*. One can straightforwardly obtain the equation of motion for the scalar field by taking the trace of Eq. 5.2.2:

$$\Box f_R = \frac{1}{3}\left(R - f_R R + 2f(R) + 8\pi G\rho_m\right), \tag{5.2.3}$$

in which ρ_m is the matter density in the Universe. Since we are interested in the cosmological properties of these models, we need to derive the perturbation equations. In order to do this, we will work in the Newtonian gauge:

$$\mathrm{d}s^2 = (1 + 2\Psi)\mathrm{d}t^2 - a^2(t)(1 - 2\Phi)\mathrm{d}\mathbf{x}^2, \tag{5.2.4}$$

where Ψ and Φ are the gravitational potentials, with $\Psi \neq \Phi$ for the time being (non no-slip condition), t is the physical time, $\mathbf{x}$ is the comoving coordinate, and a is cosmic scale factor, with $a = 1$ today. The perturbation is around the standard Friedmann–Robertson–Walker (FRW) metric, which describes the background evolution of the Universe (or of $a(t)$). Given this, we can then write down the scalaron equation of motion:

$$\frac{1}{a^2} \nabla^2 f_R \approx -\frac{1}{3} \left[R - \bar{R} + 8\pi G \left(\rho_m - \bar{\rho}_m \right) \right] , \qquad (5.2.5)$$

and the modified Poisson equation:

$$\frac{1}{a^2} \nabla^2 \Phi \approx \frac{16\pi G}{3} \left(\rho_m - \bar{\rho}_m \right) + \frac{1}{6} \left(R - \bar{R} \right) , \qquad (5.2.6)$$

where quantities with an overbar signify those defined in the background cosmology, and ∇ denotes the three-dimensional spatial derivative with respect to $\mathbf{x}$.

When deriving Eqs. 5.2.5 and 5.2.6, we have assumed that $|f(R)| \ll |R|$ and $|f_R| \ll 1$, which is true for the models we study below. Equations 5.2.5 and 5.2.6 are solved by the standard ECOSMOG code, in which the quasi-static approximation has been used and time derivatives of the scalaron field f_R are neglected. We will show below how to extend these equations consistently to restore those time derivatives.

5.2.2   The Chameleon Screening Mechanism

While modifying the theory of gravity to explain the accelerated expansion of the Universe on a cosmological level, one must bear in mind the tremendous success of GR in Solar System tests. $f(R)$ gravity incurs a fifth force that enhances gravity on large scales, which needs to be suppressed locally to pass those experimental tests. For this reason, viable $f(R)$ models are equipped with a mechanism to ensure that: (1) gravity is modified (enhanced) on cosmological scales, and (2) GR is recovered in Solar or similar systems. This is known as the *chameleon mechanism*.

To see how this is manifest in $f(R)$ gravity, we can construct an effective potential for the scalaron field as:

$$\frac{d V_{\text{eff}} \left(f_R; \rho_m \right)}{d f_R} = -\frac{1}{3} \left[R - f_R R + 2 f(R) + 8\pi G \rho_m \right] . \qquad (5.2.7)$$

In regions of high matter density ($\rho_m \gg \bar{\rho}_m$), $|f_R| \ll |\bar{f}_R| \ll 1$, and so the GR solution $R = -8\pi G \rho_m$ minimises Eq. 5.2.7, giving rise to an effective mass for the scalaron field:

$$m_{\text{eff}}^2 \equiv \frac{d^2 V_{\text{eff}}}{d f_R^2} \approx -\frac{1}{3} \frac{dR}{d f_R} > 0 . \qquad (5.2.8)$$

This fifth force is Yukawa-type, and decays as $\exp(-m_{\text{eff}} r)$, where r is the separation between two test masses. According to Eq. 5.2.7, m_{eff} depends explicitly on ρ_m, and we can see from Eq. 5.2.8 that in regions of high matter density (or equally, where the Newtonian potential is deep), the fifth force is more strongly suppressed as m_{eff} is larger there (which is because $|R| \approx 8\pi G\rho_m$ is large and $|f_R|$ small in high-density regions). The deviations from GR become practically undetectable, and hence the GR limit is recovered in those regimes.

5.2.3 The Hu–Sawicki Model

Thus far, the discussion has been quite general, without specifying the functional form for $f(R)$. Note that the choice for the form of $f(R)$ completely specifies the model. The Hu–Sawicki model is one such example, which takes the following form:

$$f(R) = -M^2 \frac{c_1 \left(-R/M^2\right)^n}{c_2 \left(R/M^2\right)^n + 1}, \tag{5.2.9}$$

where M is a characteristic mass scale, defined by $M^2 = 8\pi G\bar{\rho}_{m0}/3 = H_0^2 \Omega_m$, with $\bar{\rho}_{m0}$ being the background matter density today, and Ω_m the present-day fractional energy density of matter. H_0 is the Hubble expansion rate today. c_1, c_2 and n are free parameters of the theory. One can then show that:

$$f_R = -\frac{c_1}{c_2^2} \frac{n \left(-R/M^2\right)^{n-1}}{\left[\left(-R/M^2\right)^n + 1\right]^2}. \tag{5.2.10}$$

Given that:

$$-\bar{R} \approx 8\pi G\bar{\rho}_m - 2\bar{f}(R) = 3M^2 \left[a^{-3} + \frac{2}{3}\frac{c_1}{c_2}\right], \tag{5.2.11}$$

in order to match the ΛCDM background expansion, we set $c_1/c_2 = 6\Omega_\Lambda/\Omega_m$. In this chapter, we use $\Omega_m = 0.281$ and $\Omega_\Lambda \equiv 1 - \Omega_m = 0.719$ from WMAP9 (Hinshaw et al. 2013). In doing so, we find that $-\bar{R} \approx 34M^2 \gg M^2$, so that we can further simplify Eq. 5.2.10 as:

$$f_R \approx -n\frac{c_1}{c_2^2} \left[\frac{M^2}{-R}\right]^{n+1}. \tag{5.2.12}$$

Finally, we define $\xi \equiv c_1/c_2^2$, and essentially reduce the Hu–Sawicki model into a two-parameter family in (n, ξ). This is because once the background evolution is fixed to match that of ΛCDM as a good approximation, it is only the combination c_1/c_2^2 that appears in the $f(R)$ field equations.

5.3 $f(R)$ Equations

5.3.1   The Newtonian-Gauge Perturbation Variables

In what follows, we shall work in the Newtonian gauge, defined in Eq. 5.2.4. With the usual definitions of the Christoffel coefficients and the Ricci tensor as:

$$\Gamma^\gamma_{\alpha\beta} = \frac{1}{2} g^{\gamma\eta} \left(\partial_\beta g_{\alpha\eta} + \partial_\alpha g_{\beta\eta} - \partial_\eta g_{\alpha\beta} \right), \text{ and} \tag{5.3.13}$$

$$R_{\mu\nu} = \partial_\gamma \Gamma^\gamma_{\mu\nu} - \partial_\nu \Gamma^\gamma_{\mu\gamma} + \Gamma^\lambda_{\gamma\lambda} \Gamma^\gamma_{\mu\nu} - \Gamma^\lambda_{\gamma\mu} \Gamma^\gamma_{\lambda\nu}, \tag{5.3.14}$$

where ∂_α is the partial derivative with respect to x^α, we find, up to first order in perturbation variables Φ and Ψ,

$$\begin{aligned}
\Gamma^0_{00} &\approx \dot{\Psi}, \\
\Gamma^0_{0i} &\approx \partial_i \Psi, \\
\Gamma^i_{00} &\approx \frac{1}{a^2} \delta^{ij} \partial_j \Psi, \\
\Gamma^i_{j0} &\approx \left(H - \dot{\Phi} \right) \delta^i_j, \\
\Gamma^0_{ij} &\approx a^2 H \left(1 - 2\Psi - 2\Phi \right) \delta_{ij} - a^2 \dot{\Phi} \delta_{ij}, \\
\Gamma^i_{jk} &\approx -\partial_k \Phi \delta^i_j - \partial_j \Phi \delta^i_k + \partial^i \Phi \delta_{jk},
\end{aligned} \tag{5.3.15}$$

where the overdots indicate derivatives with respect to the physical time t, and $H = \dot{a}/a$. The corresponding Ricci tensor components are:

$$\begin{aligned}
R_{00} &\approx \frac{1}{a^2} \Psi^{,i}_{,i} - 3 \left(\dot{H} + H^2 \right) \\
&\quad +3\ddot{\Phi} + 3H \left(\dot{\Psi} + 2\dot{\Phi} \right),
\end{aligned} \tag{5.3.16}$$

$$R_{0i} \approx 2\dot{\Phi}_{,i} + 2H\Psi_{,i}, \tag{5.3.17}$$

$$\begin{aligned}
R_{ij} &\approx (\Phi - \Psi)_{,ij} + \Phi^{,k}_{,k} \delta_{ij} - a^2 \ddot{\Phi} \delta_{ij} \\
&\quad +a^2 \left(\dot{H} + 3H^2 \right) (1 - 2\Phi - 2\Psi) \delta_{ij} \\
&\quad -a^2 H \left(\dot{\Psi} + 6\dot{\Phi} \right) \delta_{ij}.
\end{aligned} \tag{5.3.18}$$

By using the definition of the Ricci scalar:

$$R = g^{\mu\nu} R_{\mu\nu}, \tag{5.3.19}$$

in conjunction with Eq. 5.2.4, we obtain:

$$R \approx \frac{1}{a^2} \left(2\Psi^{,i}_{,i} - 4\Phi^{,i}_{,i} \right) + 6\ddot{\Phi} + 6H \left(\dot{\Psi} + 4\dot{\Phi} \right)$$
$$- 6 \left(\dot{H} + 2H^2 \right) (1 - 2\Psi) \ . \tag{5.3.20}$$

Finally, with the definition of the Einstein tensor as:

$$G^{\mu}_{\nu} = R^{\mu}_{\nu} - \frac{1}{2} \delta^{\mu}_{\nu} R \ , \tag{5.3.21}$$

we find:

$$G^0_0 \approx \frac{2}{a^2} \Phi^{,i}_{,i} + 3H^2 - 6H \left(\dot{\Phi} + H\Psi \right) \ ,$$
$$G^0_i \approx 2\dot{\Phi}_{,i} + 2H\Psi_{,i} \ ,$$
$$G_{ij} \approx (\Phi - \Psi)_{,ij} + (\Psi - \Phi)^{,k}_{,k} \delta_{ij} + 3a^2 \ddot{\Phi} \delta_{ij}$$
$$2a^2 H^2 (1 - 2\Psi) \delta_{ij} - 3H^2 a^2 (1 - 2\Psi) \delta_{ij}$$
$$+ a^2 H \left(2\dot{\Psi} + 6\dot{\Phi} \right) \delta_{ij} \ . \tag{5.3.22}$$

5.3.2 The Modified $f(R)$ Equation of Motion

The scalaron equation of motion (Eq. 5.2.5) assumes the quasi-static approximation (i.e., the time derivatives of the scalaron field are neglected), and hence needs to be generalised for the study here. We therefore re-derive the equation of motion in the Newtonian gauge using Eqs. 5.3.13–5.3.22. Using the definition that $\Box f_R = g^{\mu\nu} \nabla_\mu \nabla_\nu f_R$, we find that in the Newtonian gauge:

$$\Box f_R = (1 - 2\Psi) \ddot{f}_R - \frac{1}{a^2} f^{,i}_{R,i}$$
$$+ \left[3H (1 - 2\Psi) - \dot{\Psi} - 2\dot{\Phi} \right] \dot{f}_R \ . \tag{5.3.23}$$

When deriving Eq. 5.3.23, we have retained terms involving $\Psi \dot{\Phi}, \dot{\Psi}$, but neglect second-order terms such as $\Phi^{,i} \Phi_{,i}$ and $\Phi\dot{\Phi}$. In what follows, we also make use of the following relations:

$$|\Phi| \sim |\Psi| \ll 1, \quad |f_R| \ll 1, \quad |\dot{\Phi}| \sim |\dot{\Psi}| \ll H,$$
$$|\ddot{\Phi}| \sim |\ddot{\Psi}| \sim H|\dot{\Psi}| \sim H|\dot{\Phi}| \ll H^2 \sim |\dot{H}| \ , \tag{5.3.24}$$

so that quantities on the left-hand side of the inequalities can be neglected when compared to the terms on the right-hand sides.

Since we are interested in the effects of the field perturbations, we need to subtract the contribution of the background quantities from these equations. Denoting such quantities with an overbar, and using Eq. 5.2.3, we write the following background equation of motion for the scalaron:

$$\ddot{\bar{f}}_R + 3H\dot{\bar{f}}_R \approx \frac{1}{3}\left[\bar{R} - \bar{f}_R\bar{R} + 2\bar{f}_R(R) + 8\pi G\bar{\rho}_m\right]$$
$$\approx 0, \tag{5.3.25}$$

where the second equality comes from the assumption that, at the background level, the scalaron field $\bar{f}_R$ always follows the minimum of its effective potential. In reality $\bar{f}_R$ oscillates quickly around the minimum because $m_{\text{eff}}^2 \gg H^2$, such that over many oscillations the above assumption describes the average effect well (we will revisit to this point below). Under this assumption, and because the value of the scalaron itself is quite small ($|\bar{f}_R| \leq 10^{-4}$ in the models studies here), we can assume that $|\bar{f}_R\bar{R}| \ll |\bar{R}|$, and rewrite Eq. 5.3.25 as:

$$-\bar{R} = 8\pi G\bar{\rho}_m + 2\bar{f}_R(R)$$
$$\approx 8\pi G\bar{\rho}_m + 32\pi G\bar{\rho}_\Lambda$$
$$= 8\pi G T^\mu_\mu, \tag{5.3.26}$$

where we have used the fact that when $|R| \gg M^2$, which always holds for the models studied here, $f(R)$ remains approximately constant throughout the cosmic history (cf. Eq. 5.2.9). Note that the fact that $f(R)$ remains approximately constant for different values of R means also that its perturbations are small and can be neglected, namely:

$$f(R) - \bar{f}(R) \sim f_R R - \bar{f}_R\bar{R} \ll R - \bar{R}. \tag{5.3.27}$$

Subtracting off the background part from the scalaron equation of motion, and denoting the perturbed quantities as $R - \bar{R} \equiv \delta R$ and $\rho_m - \bar{\rho}_m \equiv \delta\rho_m$, we find:

$$\ddot{f}_R + 3H\dot{f}_R - \frac{1}{a^2}\nabla^2 f_R \approx \frac{1}{3}[\delta R + 8\pi G\delta\rho_m]. \tag{5.3.28}$$

Note that the use of Eq. 5.3.25 implicates that it is $\ddot{f}_R$ that appears in this equation, rather than $\delta\ddot{f}_R$. This is convenient because later we will write δR as a function of f_R instead of $\delta f_R = f_R - \bar{f}_R$.

A quick comparison to the quasi-static version of the $f(R)$ equation of motion (Eq. 5.2.5) shows that the first two terms on the left-hand side of Eq. 5.3.28 are the additional terms one is left with when keeping the time derivatives in the scalar field equation of motion.

5.3.3 The Modified Poisson Equation

Equation 5.3.28 is one of the two equations that govern the formation of structure – the other is the modified Poisson equation. The full Einstein field equations in $f(R)$ gravity become:

$$(1 + f_R) G_{\mu\nu} = 8\pi G T_{\mu\nu} + \left[\frac{1}{2}f(R) - \frac{1}{2}f_R R - \Box f_R\right] g_{\mu\nu}$$
$$+ \nabla_\mu \nabla_\nu f_R , \quad (5.3.29)$$

with the following individual space-time components written in the Newtonian gauge:

$$\frac{2}{a^2}\Phi^{,i}_{,i} + 3H^2 \approx \frac{16}{3}\pi G\rho_m - \frac{1}{3}R - \frac{1}{6}f(R) + \ddot{f}_R \quad \text{(0−0 component, full)},$$

$$\frac{2}{a^2}\Phi^{,i}_{,i} \approx \frac{16}{3}\pi G\delta\rho_m - \frac{1}{3}\delta R + \ddot{f}_R \quad \text{(0−0 component, excluding background)},$$

$$\frac{1}{a^2}(\Psi - \Phi)^{,i}_{,j} + \delta^i_j\left[2\dot{H} + 3H^2 - \frac{1}{a^2}(\Psi - \Phi)^{,k}_{,k}\right] \approx 8\pi G T^i_j - \quad (5.3.30)$$

$$\frac{8}{3}\pi G\rho_m\delta^i_j - \frac{1}{3}R\delta^i_j - \frac{1}{6}f(R)\delta^i_j - \frac{1}{a^2}f^{,i}_{R\,,j} + \dot{f}_R H\delta^i_j$$
$$(i - j \text{ components, full}),$$

$$\frac{2}{a^2}(\Psi - \Phi)^{,i}_{,i} - 9H^2 + 6\dot{H} \approx 8\pi G(\rho_m + 3p_m) + R + \frac{1}{2}f(R) + \frac{1}{a^2}f^{,i}_{R\,,i} - 3H\dot{f}_R$$
$$\text{(Trace of } i - j \text{ components, including background)},$$

$$\frac{2}{a^2}(\Psi - \Phi)^{,i}_{,i} \approx 8\pi G(\rho_m + 3p_m) - 8\pi G(\bar{\rho}_m + 3\bar{p}_m) + \delta R + \frac{1}{a^2}f^{,i}_{R\,,i} - 3H\dot{f}_R$$
$$\text{(Trace of } i - j \text{ components, excluding background)}. \quad (5.3.31)$$

In the above, the equations marked as 'excluding background' are obtained by directly subtracting the ΛCDM background Friedmann equations from the full (00) and (ij) components of the modified Einstein equations, and using $\bar{f}(R) \approx 16\pi G\rho_\Lambda$ (cf. Eq. 5.3.26). This is why terms such as $\ddot{f}_R$ and $3H\dot{f}_R$ appear in them, rather than $\ddot{f}_R - \ddot{\bar{f}}_R$ and $3H\dot{f}_R - 3H\dot{\bar{f}}_R$.

The Poisson equation can be obtained by taking the trace of the Einstein field equation, and from this we get:

$$\frac{1}{a^2}\nabla^2\Psi \approx \frac{16\pi G}{3}\delta\rho_m + \frac{1}{6}\delta R + \ddot{f}_R , \quad (5.3.32)$$

where Eq. 5.3.28 has been used to eliminate $\frac{1}{a^2}f^{,i}_{R\,,i} - 3H\dot{f}_R$ in Eq. 5.3.30.

Equation 5.3.32 alongside Eq. 5.3.28 are the two that we need to solve and use to update the simulation particle positions to quantify the effect of non-vanishing time derivatives of f_R.

We would like to make a final note before concluding this section. In principle, terms such as $H\dot{\Psi}$ can be of the order of $H\dot{f}_R$, even though we have neglected them here. In our investigation, however, the aim is not to numerically solve all possible non-static terms, but rather to consistently investigate the effects of terms in $\dot{f}_R$ and $\ddot{f}_R$. Therefore, even though our equations are in some sense incomplete, they are sufficient for our specific purpose here.

5.4 Evolution Equations in ECOSMOG

Our N-body simulations are performed using the massively-parallelised ECOSMOG code (Li et al. 2012), which is based on the adaptive mesh refinement (AMR) code RAMSES (Teyssier 2002). An AMR code can resolve high-density regions by refining (i.e., splitting) a mesh cell into eight sub-cells, when the number of particles within it exceeds some predefined threshold. This is particularly useful in $f(R)$ gravity simulations, where it is necessary in the high-density regions to achieve adequate resolution in order to solve the non-linear field equations and accurately quantify the chameleon effect. The code employs a multigrid relaxation algorithm, arranged in V-cycles (i.e., alternating between coarse and fine meshes to solve the field equations), to accelerate the convergence of the solution (Press et al. 2002).

5.4.1 Equations in Code Units

In order to solve Eqs. 5.3.28 and 5.3.32, we need to convert the quantities in those equations to the superconformal units used by ECOSMOG, summarised in the equations below:

$$\tilde{x} = \frac{x}{B}, \quad \tilde{\rho} = \frac{\rho a^3}{\rho_c \Omega_m}, \quad \tilde{v} = \frac{av}{B H_0},$$

$$\tilde{\Psi} = \frac{a^2 \Psi}{(B H_0)^2}, \quad d\tilde{t} = H_0 \frac{dt}{a^2}, \quad \tilde{c} = \frac{c}{B H_0},$$

$$\tilde{f}_R = a^2 f_R . \tag{5.4.33}$$

Here, x is the comoving coordinate, a is the scale factor, ρ_c the critical density of the Universe today, v is the particle velocity, Ψ is the gravitational potential and c is the speed of light. Furthermore, B is the comoving size of the simulation box in units of h^{-1} Mpc, whereas $H_0 = 100\,h\,\mathrm{km s^{-1}\,Mpc^{-1}}$. Under these conventions, the new terms appearing from the inclusion of the time derivatives become:

$$\dot{f}_R = a^{-2} \dot{\tilde{f}}_R - 2a^{-2} H \tilde{f}_R$$

$$\ddot{f}_R = a^{-2} \ddot{\tilde{f}}_R - 4a^{-2} H \dot{\tilde{f}}_R - 2a^{-2} \dot{H} \tilde{f}_R$$

$$+ 4a^{-2} H^2 \tilde{f}_R . \tag{5.4.34}$$

For the Hu–Sawicki model, this then transforms the modified Poisson equation (Eq. 5.3.32) and the $f(R)$ equation of motion (Eq. 5.3.28) into:

$$\tilde{\nabla}^2 \tilde{\Psi} = 2\Omega_m a \left(\tilde{\rho} - 1\right) + \frac{1}{6}\Omega_m a^4 \left[\left(\frac{-na^2\xi}{\tilde{f}_R}\right)^{\frac{1}{n+1}} - 3\left(a^{-3} + 4\frac{\Omega_\Lambda}{\Omega_m}\right)\right] \quad (5.4.35)$$

$$+ \left[a^{-2}\frac{d^2 \tilde{f}}{d\tilde{t}^2} - 6\frac{H}{H_0}\frac{d\tilde{f}}{d\tilde{t}} + 2a^2 \left(2\frac{H^2}{H_0^2} - \frac{\dot{H}}{H_0^2}\right)\tilde{f}_R\right],$$

$$\tilde{\nabla}^2 \tilde{f}_R = \frac{-1}{\tilde{c}^2}\Omega_m a \left(\tilde{\rho} - 1\right) - \frac{1}{3\tilde{c}^2}\Omega_m a^4 \left[\left(\frac{-na^2\xi}{\tilde{f}_R}\right)^{\frac{1}{n+1}} - 3\left(a^{-3} + 4\frac{\Omega_\Lambda}{\Omega_m}\right)\right]$$

$$+ \frac{1}{\tilde{c}^2}\left[a^{-2}\frac{d^2 \tilde{f}}{d\tilde{t}^2} - 3\frac{H}{H_0}\frac{d\tilde{f}}{d\tilde{t}} - 2a^2 \left(\frac{H^2}{H_0^2} + \frac{\dot{H}}{H_0^2}\right)\tilde{f}_R\right] \quad (5.4.36)$$

Note that all terms in the above equations are dimensionless (dimensional quantities, such as H and $\dot{H}$, are properly normalised using H_0). We have also carefully distinguished between overdots (derivatives with respect to the physical time t) and $d/d\tilde{t}$ (derivatives with respect to the superconformal time $\tilde{t}$), such that the former only applies to purely background quantities such as a and H. Since the background evolution is approximated in the same way as in ΛCDM, quantities such as H/H_0 and $\dot{H}/H_0^2$ can be obtained analytically.

5.4.2 Discretising the Equations

In this section, we discretise the equations in Eq. 5.4.35 to make them appropriate for implementation in ECOSMOG. During its time evolution, the value of $\tilde{f}_R$ can be very close to zero, and to avoid numerical problems, we solve for a different variable, $\tilde{f}_R \equiv -e^{\tilde{u}}$, instead. The current value of a quantity ϕ in the grid cell (i, j, k) will be identified as $\phi_{i,j,k}$. Since everything presented below is already in the code units, we will drop tilde symbols in the discretised Poisson and $f(R)$ equations for clarity wherever this will not cause confusion (we keep the tilde in $\tilde{c}$, however). Given a cell size h, we obtain for the Poisson equation:

$$\frac{1}{h^2}\left[\Psi_{i+1,j,k} + \Psi_{i-1,j,k} + \Psi_{i,j+1,k} + \Psi_{i,j-1,k} + \Psi_{i,j,k+1} + \Psi_{i,j,k-1} - 6\Psi_{i,j,k}\right]$$

$$= 2\Omega_m a \left(\rho_{i,j,k} - 1\right) - \frac{1}{6}\Omega_m a^4 \left[(na^2\xi)^{\frac{1}{n+1}}\exp\left(\frac{-u_{i,j,k}}{n+1}\right) - 3\left(a^{-3} + 4\frac{\Omega_\Lambda}{\Omega_m}\right)\right]$$

$$+ \left[a^{-2}\Delta t^{-1}\left[d f dt_{i,j,k}^{(n)} - d f dt_{i,j,k}^{(n-1)}\right] + 6\frac{H}{H_0}\Delta t^{-1}\exp\left(u_{i,j,k}\right)\left(u_{i,j,k} - u_{i,j,k}^{(n-1)}\right)\right.$$

$$\left. -2a^2 \left(2\frac{H^2}{H_0^2} - \frac{\dot{H}}{H_0^2}\right)\exp\left(u_{i,j,k}\right)\right], \quad (5.4.37)$$

where Δt is the time step in code units adopted by the simulation. The last line in Eq. 5.4.37 contains the additional terms that arise from going beyond the quasi-static approximation.

Discretising the $f(R)$ equation of motion is a similar, if slightly more laborious task. In order to reduce clutter, we define a variable $b \equiv e^u$, and write the discrete scalaron equation as:

$$
\begin{aligned}
&\frac{1}{h^2} \left[b_{i+\frac{1}{2},j,k} u_{i+1,j,k} - u_{i,j,k} \left(b_{i+\frac{1}{2},j,k} + b_{i-\frac{1}{2},j,k} \right) + b_{i-\frac{1}{2},j,k} u_{i-1,j,k} \right] \\
&+ \frac{1}{h^2} \left[b_{i,j+\frac{1}{2},k} u_{i,j+1,k} - u_{i,j,k} \left(b_{i,j+\frac{1}{2},k} + b_{i,j-\frac{1}{2},k} \right) + b_{i,j-\frac{1}{2},k} u_{i,j-1,k} \right] \\
&+ \frac{1}{h^2} \left[b_{i,j,k+\frac{1}{2}} u_{i,j,k+1} - u_{i,j,k} \left(b_{i,j,k+\frac{1}{2}} + b_{i,j,k-\frac{1}{2}} \right) + b_{i,j,k-\frac{1}{2}} u_{i,j,k-1} \right] \\
&+ \frac{1}{3\bar{c}^2} \Omega_m a^4 \left(na^2 \xi \right)^{\frac{1}{n+1}} \exp \left(-\frac{u_{i,j,k}}{n+1} \right) - \frac{1}{\bar{c}^2} \Omega_m a \left(\delta_{i,j,k} - 1 \right) - \frac{1}{\bar{c}^2} \Omega_m a^4 \left(a^{-3} + 4\frac{\Omega_\Lambda}{\Omega_m} \right) \\
&+ \frac{1}{\bar{c}^2} \left[-a^{-2} \Delta t^{-1} \mathrm{d}f \mathrm{d}t_{i,j,k}^{(n-1)} - a^{-2} \Delta t^{-2} \exp \left(u_{i,j,k} \right) \left(u_{i,j,k} - u_{i,j,k}^{(n-1)} \right) \right. \\
&\left. + 3\frac{H}{H_0} \Delta t^{-1} \exp \left(u_{i,j,k} \right) \left(u_{i,j,k}^{(n)} - u_{i,j,k}^{(n-1)} \right) + 2a^2 \left(\frac{\dot{H}}{H_0^2} + \frac{H^2}{H_0^2} \right) \exp \left(u_{i,j,k} \right) \right] \qquad (5.4.38) \\
&= 0.
\end{aligned}
$$

Once again, the effect of the time derivatives is incorporated in the terms in the last two lines of Eq. 5.4.38. Taking the second order derivative with respect to the x coordinate as an example, this scheme gives:

$$
\frac{\partial^2 \phi}{\partial x^2} \rightarrow \frac{1}{h^2} \left(\phi_{i+1,j,k} - 2\phi_{i,j,k} + \phi_{i-1,j,k} \right),
$$

where h is the size of the mesh cell and the subscript i,j,k refers to the cell that is ith in the x direction, jth in the y direction and kth in the z direction. Note that the discrete Laplacian in Eq. (5.4.38) looks slightly more complicated because $\tilde{\nabla}^2 e^u \equiv \tilde{\nabla} \cdot \left(e^u \tilde{\nabla} u \right)$, and we have defined $b \equiv e^u$ such that:

$$
b_{i+\frac{1}{2},j,k} \equiv \frac{1}{2} \left[\exp \left(u_{i+1,j,k} \right) + \exp \left(u_{i,j,k} \right) \right],
$$

$$
b_{i-\frac{1}{2},j,k} \equiv \frac{1}{2} \left[\exp \left(u_{i-1,j,k} \right) + \exp \left(u_{i,j,k} \right) \right].
$$

We have seen in Eq. 5.4.35 that their discrete versions will contain terms like $\mathrm{d}^2 \tilde{f}/\mathrm{d}\tilde{t}^2$ and $\mathrm{d}\tilde{f}/\mathrm{d}\tilde{t}$ in our code units. By discretising also in time, we find that:

$$\frac{d\tilde{f}}{d\tilde{t}} = -\Delta t^{-1} \exp\left(u_{i,j,k}\right)\left(u_{i,j,k}^{(n)} - u_{i,j,k}^{(n-1)}\right)$$

$$\equiv df dt_{i,j,k}^{(n)},$$

$$\frac{d^2\tilde{f}}{dt^2} = \Delta t^{-1}\left[df dt_{i,j,k}^{(n)} - df dt_{i,j,k}^{(n-1)}\right], \tag{5.4.39}$$

in which $u_{i,j,k}^{(n)}$ and $u_{i,j,k}^{(n-1)}$ are respectively the values of the scalaron field in the current time step (n) and the previous time step $(n-1)$. Throughout this chapter $u_{i,j,k}$ without a superscript (n) always denotes the value at step (n). In the simulations, the code records $u_{i,j,k}^{(n)}$ and $df dt_{i,j,k}^{(n)}$ for each cell so that in the step that follows, they can be used as $u_{i,j,k}^{(n-1)}$ and $df dt_{i,j,k}^{(n-1)}$. Note that in principle we also need the value $u_{i,j,k}^{(n-2)}$ to evaluate $d^2 f/dt^2$ at step (n), but in practice this is implicitly included in the calculation of $df dt_{i,j,k}^{(n-1)}$ at step $(n-1)$.

By doing the above, we are incorporating the time derivatives in an *implicit* way, in contrast to the *explicit* method that tries to evolve the scalar field by:

$$u_{i,j,k}^{(n)} = u_{i,j,k}^{(n-1)} + \frac{d}{dt}u_{i,j,k}^{(n-1)}\Delta t. \tag{5.4.40}$$

It is known that the implicit scheme of numerical integration is usually more stable than the explicit method. However, the main advantage of our method is that it does not change the property that the $f(R)$ equation, Eq. 5.4.38, is a boundary-value problem and therefore can be solved using a relaxation algorithm, with very little change to the code structure of ECOSMOG. The explicit scheme described in Eq. 5.4.40, on the other hand, means that the equation becomes an initial-value problem. Of course, because we are evaluating the time derivatives in a 'backward' manner (that is, we are computing $du^{(n)}/dt$ and $d^2u^{(n)}/dt^2$ by using $u^{(n-1)}$ and $u^{(n-2)}$ rather than using variables evaluated at step (n)), this will inevitably introduce numerical errors in evolving the differential equation. However, by making the time steps short enough, the two methods should agree, and therefore a consistency check can always be done by reducing Δt to confirm that the method works properly, as we will demonstrate below.

Another important point needs to be made at this stage. As mentioned above, the value of the scalar field, f_R, oscillates quickly around the local potential minimum. Therefore, in order to calculate its time evolution, our procedure in Eq. 5.4.39 implicitly performs an average over the many oscillations in each time step of the simulation. Evaluating a more "instantaneous" time derivative accurately would require a huge number of time steps, especially in high-density regions, where the scalaron mass m_{eff} is larger and so the scalar field oscillates faster, and is therefore not practical for our $f(R)$ simulations. For linear terms, such as $\ddot{f}_R$ and $H\dot{f}_R$, the order of doing the time average and solving the scalaron equation can be freely swapped, and therefore the procedure in Eq. 5.4.39 is expected to work without any problem. On the other hand, for the non-linear terms in the scalaron equation, such as $\delta R(f_R)$, the order does matter, and using time-averaged values for f_R will introduce errors which are expected to become larger if the non-linearity gets stronger. For our simulations, however, we do not expect such errors to be significant enough to affect our conclusions; we will revisit and quantify this point in Sect. 5.6.3.

Equation 5.4.38 can be thought of symbolically as an equation involving a non-linear differential operator, in the form:

$$\mathscr{L}^h u_{i,j,k} = 0, \qquad (5.4.41)$$

where the superscript h indicates that the operator is acting on a level where the cell size is h. The Gauss–Seidel relaxation in ECOSMOG then updates the scalar field as:

$$u_{i,j,k}^{h,\,(\text{new})} = u_{i,j,k}^{h,\,(\text{old})} - \frac{\mathscr{L}^h\left(u_{i,j,k}^{h,\,(\text{old})}\right)}{\frac{\partial \mathscr{L}^h\left(u_{i,j,k}^{h,\,(\text{old})}\right)}{\partial u_{i,j,k}^{h,\,(\text{old})}}}. \qquad (5.4.42)$$

The form of the denominator in the above equation is given by:

$$\frac{\partial \mathscr{L}^h(u_{i,j,k})}{\partial u_{i,j,k}}$$

$$= \frac{\tilde{c}^2}{2h^2} b_{i,j,k} \left[u_{i+1,j,k} + u_{i-1,j,k} + u_{i,j+1,k} + u_{i,j-1,k} + u_{i,j,k+1} + u_{i,j,k-1} - 6u_{i,j,k} \right]$$

$$- \frac{\tilde{c}^2}{2h^2} \left[b_{i+1,j,k} + b_{i-1,j,k} + b_{i,j+1,k} + b_{i,j-1,k} + b_{i,j,k+1} + b_{i,j,k-1} + 6b_{i,j,k} \right]$$

$$- \frac{1}{3(n+1)} \Omega_m a^4 \left(na^2\xi\right)^{\frac{1}{n+1}} \exp\left(-\frac{u_{i,j,k}}{n+1}\right)$$

$$+ \frac{1}{\tilde{c}^2} \left[\left(3\frac{H}{H_0}\Delta t^{-1} - a^{-2}\Delta t^{-2} \right) \exp\left(u_{i,j,k}\right)\left(1 + u_{i,j,k} - u_{i,j,k}^{(n-1)}\right) \right. \qquad (5.4.43)$$

$$\left. + 2a^2 \left(\frac{\dot{H}}{H_0^2} + \frac{H^2}{H_0^2} \right) \exp\left(u_{i,j,k}\right) \right].$$

Again, in the above equation, the last line represents the additional terms that arise from the inclusion of the time derivatives, while the first three lines are exactly the same as in the ordinary quasi-static case.

For more details about how the above discrete equations are implemented in ECOSMOG and the associated technical details, such as the boundary conditions, the interested readers are referred to the original ECOSMOG code paper (Li et al. 2012).

5.4.3 Time Integration

Since the main goal of our chapter is to assess the importance of time derivatives in $f(R)$ simulations, the choice of time step is of fundamental importance. In ECOSMOG, this is determined using the Courant–Friedrichs–Lewy (CFL) condition (Courant et al. 1928; Li et al. 2012), which is required for the stability of numerical integrations. In our simulations, this condition essentially requires that the size of a physical time step dt has to be smaller than the time it takes for a particle to travel to an adjacent grid cell. Denoting the particle velocity by v, and the physical size of a cell in the grid as dx, then the CFL condition dictates that in a particular time step these quantities are linked by:

$$\frac{1}{v}\frac{dx}{dt} \geq \mathcal{O}(1) . \qquad (5.4.44)$$

This condition must be satisfied at every time step for the solution to be stable. Using that $v \ll c$, in code units (Eq. 5.4.33), this condition translates to:

$$\frac{d\tilde{x}^2}{a^2\tilde{c}^2 d\tilde{t}^2} \ll 1 . \qquad (5.4.45)$$

Recall (Eq. 5.4.33) that $\tilde{c} = c/BH_0$, where B is the box size of the simulation. For a fixed box size, Eq. 5.4.45 then tells us that:

$$h^2 \ll a^2\tilde{c}^2 \Delta t^2, \qquad (5.4.46)$$

where $h = d\tilde{x}$ and $\Delta t = d\tilde{t}$.

The above equation already hints at the answer to our question regarding the importance of time derivatives relative to spatial derivatives. It tells us that the manner in which the size of the time step (Δt) is set is such that it is generally much larger (with a multiplicative factor of $a\tilde{c}$) than the size of the cell (h). As a result, in the discrete scalaron equation above (Eq. (5.4.38)), one would expect the spatial variation of the scalar field (terms proportional to h^{-2}) to be more significant than its variation in time (terms proportional to $(a\tilde{c}\Delta t)^{-2}$) – or, in other words, that a quasi-static approximation is good. In the following section, we will proceed to perform N-body simulations to confirm our expectation from these simple order-of-magnitude arguments.

5.5 Results

In this section, we apply our modified ECOSMOG code to perform N-body simulations of the Hu–Sawicki model, for three different choices of the present-day value of $\bar{f}_R$, namely $|\bar{f}_{R0}| = 10^{-4}, 10^{-5}, 10^{-6}$, which we will refer to as F4, F5 and F6 respec-

Table 5.1 Summary of simulations performed in this work

Name	Model	L_{box} (h^{-1} Mpc)	Particles	Realisations
L_{256}	ΛCDM, F4, F5, F6	256	256^3	5
$L_{256/2}$	ΛCDM, F6	256	256^3	1
L_{128}	ΛCDM, F6	128	256^3	5
L_{64}	ΛCDM, F6	64	256^3	1

tively. F4 (F6) forms an upper (lower) bound for cosmologically interesting $f(R)$ models: for $\left| \bar{f}_{R0} \right| > 10^{-4}$, the models are unlikely to satisfy local gravity constraints in the Milky Way (Schmidt et al. 2009b), whereas for $\left| \bar{f}_{R0} \right| < 10^{-6}$, the differences from GR are very small. In what follows, we also set the parameter $n = 1$ (Eq. 5.2.9).

The cosmological parameters for our N-body simulations are the same as in the best-fitting WMAP9 cosmology (Hinshaw et al. 2013), with $\Omega_m = 0.281$, $\Omega_\Lambda = 0.719$, $h = 0.697$, $n_s = 0.972$ and $\sigma_8 = 0.82$. Here, $h = H_0/(100 \text{ km/s/Mpc})$ is the dimensionless Hubble parameter, n_s is the spectral index of the primordial power spectrum, and σ_8 is the linear rms density fluctuation in a sphere of radius 8 h^{-1} Mpc, at $z = 0$. We expect that our findings here should not change with a different choice of parameters.

We perform low-resolution runs for ΛCDM, F4, F5 and F6 in a box of size 256 h^{-1} Mpc with 256^3 particles, and higher-resolution runs for ΛCDM and F6 in a box size of 128 h^{-1} Mpc with 256^3 particles. In each case, we simulate 5 realisations of initial conditions (the same initial conditions are used for ΛCDM and $f(R)$ simulations because at the initial time, $z_i = 49$, the effect of modified gravity is still negligible for F6, F5 and F4). For every $f(R)$ simulation we have performed for this work, we conduct both a quasi-static run and a time derivative run, to quantify the impact of including non-static effects. To check for the influence of changing the size of the time step and of resolution, we simulate two additional models: $L_{256/2}$ and L_{64}. The former has the same parameters as the L_{256} run, but here we artificially halve the time step that the ECOSMOG code would naturally adopt. The latter constitutes our highest resolution run, with 256^3 particles within a box of 64 h^{-1} Mpc. In each set of simulations, the regular simulation mesh has 256 cells on each side, and is adaptively refined when the number of particles within a cell is greater than 8. A summary of the simulation details is given in Table 5.1.

5.5.1 The Matter and Velocity Divergence Power Spectra

As remarked on earlier, the first order difference between $f(R)$ simulations in the quasi-static approximation and the non-static limit can be seen in changes to the matter power spectrum (He et al. 2013). In $f(R)$ gravity, one would expect the scalaron field, through the fifth force it mediates (where the chameleon screening is not effective), to enhance the ordinary gravitational interaction, thereby strengthening the clustering of matter. To quantify this further, we define the dark matter density field $\rho\,(\mathbf{x}, t)$ as:

$$\rho\,(\mathbf{x}, t) = \bar{\rho}(t)[1 + \delta(\mathbf{x}, t)], \qquad (5.5.47)$$

where $\bar{\rho}$ is the background density field at time t, and δ encodes the fluctuations around that homogeneous background. In order to calculate the power spectrum, it is first convenient to rewrite the density contrast δ in Fourier space:

$$\delta_{\mathbf{k}} \equiv (2\pi)^{-3/2} \int \delta\,(\mathbf{x}, t)\, e^{-i\mathbf{k}\cdot\mathbf{x}} \mathrm{d}^3\mathbf{x} \,. \qquad (5.5.48)$$

The matter power spectrum is then defined by:

$$P_{\delta\delta}\,(k) \equiv P\,(k) = \langle|\delta_{\mathbf{k}}|^2\rangle. \qquad (5.5.49)$$

To measure the matter power spectrum from our simulation outputs, we make use of the publicly-available POWMES code (Colombi et al. 2009), which constructs the density field of a particle distribution by estimating the Fourier modes of the distribution using a Taylor expansion of trigonometric functions. We also compute the velocity divergence power spectra from our simulations, following the approach in Li et al. (2013c). First, we define the *expansion scalar*, which is related to the divergence of the velocity field by:

$$\theta\,(\mathbf{x}, t) = \frac{1}{aH(a)} \nabla \cdot v\,(\mathbf{x}, t) \,, \qquad (5.5.50)$$

where $v\,(x, t)$ is the cosmic peculiar velocity field and $H(a)$ is the Hubble constant at epoch a. In a similar vein to the matter power spectrum, we can take the Fourier transform of the above to get:

$$\theta_{\mathbf{k}} \equiv (2\pi)^{-3/2} \int \theta\,(\mathbf{x}, t)\, e^{-i\mathbf{k}\cdot\mathbf{x}} \mathrm{d}^3\mathbf{x} \,, \qquad (5.5.51)$$

and the corresponding velocity divergence power spectrum:

$$P_{\theta\theta}\,(k) = \langle|\theta_{\mathbf{k}}|^2\rangle \,. \qquad (5.5.52)$$

The velocity field has been shown to be more sensitive than the matter field to the effects of the fifth force, so any changes due to the inclusion of time derivatives should also have a stronger signal here (Jennings et al. 2011). We measure $\theta(\mathbf{x}, t)$ from our simulation outputs by performing a Delaunay tessellation over the discrete set of points defining the configuration of our simulation, using the publicly available DTFE code (Schaap and van de Weygaert 2000; Cautun and van de Weygaert 2011). This has the advantage of calculating a volume-weighted velocity divergence field, rather than a mass-weighted one, and also circumvents the issue of empty grid cells.

5.5.1.1 Low-Resolution Tests

As a first test, we perform simulations with 256^3 particles in the L_{256} box. To see the difference between the simulation with time derivatives and that in the quasi-static limit, we measure the enhancement of the power spectrum in each case relative to ΛCDM. In what follows, we refer to the individual cases using the notation $Fx_{\{q,t\}}$, where $x = 4, 5, 6$ indicates the value of $|\bar{f}_{R0}|$, while q (t) refers to the simulation in the quasi-static limit (with the inclusion of time derivatives).

We then smooth out the intrinsic noise in the power spectrum as follows. First, we calculate the relative difference in the power spectrum of $Fx_{\{q,t\}}$ compared with ΛCDM in each set of realisations:

$$\frac{\Delta P(k; Fx_{\{q,t\}})}{P(k; \Lambda\text{CDM})} = \frac{P(k; Fx_{\{q,t\}}) - P(k; \Lambda\text{CDM})}{P(k; \Lambda\text{CDM})}. \qquad (5.5.53)$$

We then divide the values of the wavenumber k probed by the simulation into a number of bins equally spaced in $\log(k)$, and average the relative difference in each bin over all the realisations. The scatter between realisations is represented by error bars calculated using the standard deviation in each k-bin over all realisations. The relative difference is taken with respect to ΛCDM, rather than between the quasi-static and non-static runs themselves, because the residual from the latter is expected to be very small, and taking the ratios of these small differences can look larger than they intrinsically are on a plot.

The results of the above procedure in the cases for $F4_{\{q,t\}}$, $F5_{\{q,t\}}$ and $F6_{\{q,t\}}$ are shown, respectively, in Figs. 5.1, 5.2 and 5.3. Focusing first on the quasi-static (blue symbols) simulations only, we note two features consistent in F4, F5 and F6:

Firstly, the enhancement in the matter power spectrum relative to ΛCDM closely follows the predictions of linear theory at large scales, which is what one would expect. At smaller scales, linear theory over-predicts the enhancement of power in the $f(R)$ model with respect to ΛCDM, because it fails to account for the suppression of the fifth force by the chameleon mechanism and other non-linear effects. This can also be seen in Fig. 5.1 for the F4 model, which is the one that deviates most significantly from GR – it shows a better match to linear theory for $k \leq 1\,\text{h}\,\text{Mpc}^{-1}$ compared to F5 and F6, because here the chameleon mechanism is less efficient.

Secondly, we have seen quite distinct features in $\Delta P_{\delta\delta}/P_{\delta\delta}$ for the three models. The amplitude of $\Delta P_{\delta\delta}/P_{\delta\delta}$ at $z = 0$ increases from F6 to F4, which confirms that the effect of the fifth force becomes stronger as the magnitude of the scalaron field $|\bar{f}_{R0}|$ increases. F4, for example, shows a distinct peak at around $k = 1\,\text{hMpc}^{-1}$, as demonstrated in Fig. 5.1. In F5, at these scales $\Delta P_{\delta\delta}/P_{\delta\delta}$ shows a minor flattening before rising again to smaller scales. In F6, on the other hand, there are no such noticeable features, and the enhancement of the power spectrum increases all the way down to the smallest resolved scales. These features agree well with the results of Hellwing et al. (2013), and can be explained by the different efficiency of the chameleon screening in the different models.

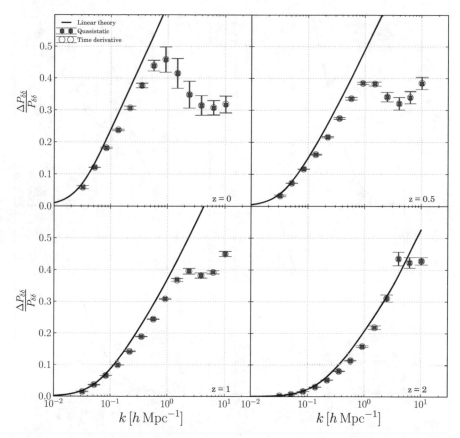

Fig. 5.1 Time evolution of excess clustering signal $\Delta P/P$ for the lower-resolution (L_{256}) F4 simulations, over four different redshifts. The open red circles represent the realisation-averaged relative difference when including the time derivatives, whereas the filled blue circles show the enhancement with respect to ΛCDM for the standard quasi-static case. The solid black line is the enhancement for quasi-static F4, relative to ΛCDM as predicted by linear theory. The procedure for calculating the averages and error bars is as described in the main text. Figure reproduced from Bose et al. 2015

A look at these figures leads us to our main result, that there is no significant change in the clustering properties when we include time derivatives into our simulations. The differences, as can be gathered from the offset between the red and blue symbols, are sub-percent. If we now look at the effect of the time derivatives (open red circles), we find that the smoothed results trace their quasi-static counterparts almost exactly. The error bars here, which represent the scatter in $\Delta P_{\delta\delta}/P_{\delta\delta}$ across realisations, almost exactly overlap as well. This is particularly true for the F4 and F5 cases, as can be seen clearly from Figs. 5.1 and 5.2. Towards smaller scales, the discrepancy between the time derivative and quasi-static runs becomes slightly more pronounced, which is because the effects of time derivatives on the fifth force will be felt at the

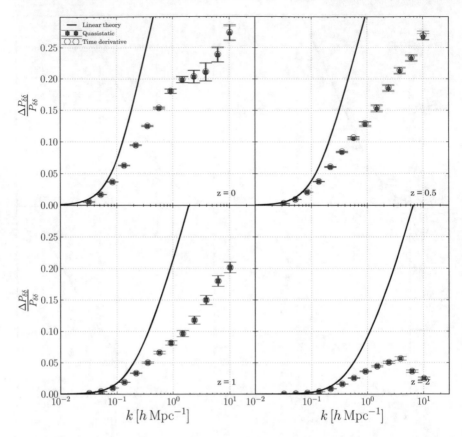

Fig. 5.2 Time evolution of $\Delta P/P$ for the lower-resolution (L_{256}) F5 simulations, over four different redshifts. The line/symbol styles are as described in Fig. 5.1. Figure reproduced from Bose et al. (2015)

smallest scales first, due to the hierarchical nature of structure formation and the properties of the initial conditions.

Inspection of Fig. 5.3 suggests that the effect of time derivatives is more significant in F6 than in F4 or F5. Here, a noticeable offset between the time derivative and quasi-static runs starts as early as $z = 0.5$. The larger effects of time derivatives could be because $\Delta P_{\delta\delta}/P_{\delta\delta}$ has a much smaller magnitude ($\leq 5\%$ down to $k \sim 10\,h\mathrm{Mpc}^{-1}$) in F6 than in F4 and F5, which makes the small impact of including the time derivatives look much stronger, but it may also arise from numerical issues (e.g., the spatial and time resolutions of our simulations are too low and the results have not yet converged). To have confidence in using our numerical simulations to do science, it is important then to understand whether this result is physical. For this reason, we need to investigate the differences between quasi-static and non-static runs when re-simulated at higher resolution. We will return to this in the next subsection.

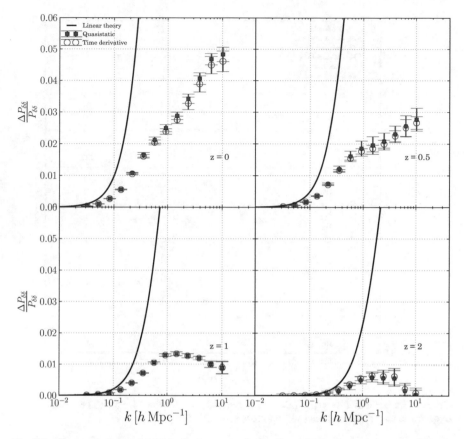

Fig. 5.3 Time evolution of $\Delta P/P$ for the lower-resolution (L_{256}) F6 simulations, over four different redshifts. Again, the line and symbol styles follow the convention in Fig. 5.1. Figure reproduced from Bose et al. (2015)

Finally, Fig. 5.4 illustrates the $z = 0$ relative difference in the velocity divergence power spectra ($\Delta P_{\theta\theta}/P_{\theta\theta}$) for F4, F5 and F6. All three models show similar features as first observed in Hellwing et al. (2013), most markedly the presence of a dip, after which the ratio $\Delta P_{\theta\theta}/P_{\theta\theta}$ increases once again. Comparison with Figs. 5.1, 5.2 and 5.3 shows that the enhancement of $\Delta P_{\theta\theta}/P_{\theta\theta}$ for these models relative to ΛCDM is a lot stronger than that in the matter power spectra – to almost an order of magnitude in the case of F6. This reiterates the aforementioned advantage of using the velocity divergence power spectrum as a more sensitive probe of modified gravity (Hellwing et al. 2014). Just as in the case of the matter power spectrum, there does not seem to be any significant difference in the enhancements when including time derivatives, as both the non- static and quasi-static simulations of the three models seem to be well-converged. Note, however, that for F6 the effects of including time derivatives on $\Delta P_{\theta\theta}/P_{\theta\theta}$ appear to be much smaller than in the case of matter power spectra, which is because of the scale on the axis.

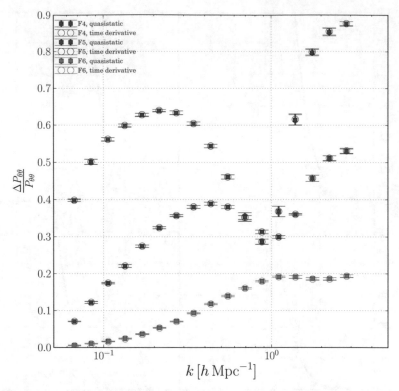

Fig. 5.4 The velocity divergence power spectrum at $z = 0$ for the F4, F5 and F6 models in the L_{256} simulation. Figure reproduced from Bose et al. (2015)

5.5.1.2 High-Resolution Tests

We have simulated the F6 model at higher resolution, by keeping the number of particles at 256^3, but using smaller boxes of size 128 and $64\,h^{-1}$ Mpc (the L_{128} and L_{64} simulations in our nomenclature). The result of the former is displayed in Fig. 5.5, from which we immediately see that the discrepancy we noticed in Fig. 5.3 is now largely reduced, even at redshift $z = 0.5$. This is demonstrated more clearly in Fig. 5.6, where in the upper panels we again plot $\Delta P_{\delta\delta}/P_{\delta\delta}$ at $z = 0$ for both the L_{256} and L_{128} runs, and show the difference between the quasi-static and non-static cases for each in the lower panel. The offset seen earlier in the L_{256} case is now essentially zero throughout all k for L_{128}, except for the smallest scales (large k) where we are likely affected by resolution once more. The case for the L_{64} simulation is shown in Fig. 5.7, but only for the snapshot at $z = 0.5$ (which shows the largest difference between the quasi-static and non-static runs in Fig. 5.5) for brevity. Again, here we see that the difference between the two is further reduced.

The implications of the results shown in Figs. 5.5 and 5.7 are twofold. First, it serves as a convergence test of our algorithm to include time derivatives in the

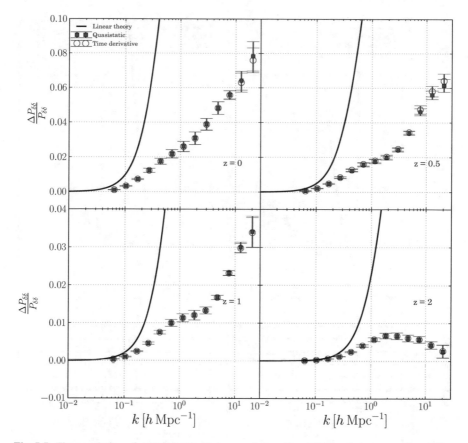

Fig. 5.5 Time evolution of $\Delta P/P$ for the higher-resolution (L_{128}) F6 simulations, over four different redshifts. Again, the line and symbol styles follow the convention in Fig. 5.1. Figure reproduced from Bose et al. (2015)

simulations and shows that, with increasing (spatial and time) resolution, the runs do converge as we anticipated. Second, it resonates our expectations and findings from F4 and F5 models, that the effect of introducing time derivatives in the F6 model has a negligible impact on the matter power spectra, compared with just the quasi-static case (if the resolution is high enough so that simulation has converged, of course).

Our conclusion is then that in all models studied in this work (which are also the most well-studied modified gravity models in the literature), the quasi-static approximation, which is adopted in almost all numerical simulations to date, is valid and is adequate to make accurate predictions for the matter and velocity divergence power spectra.

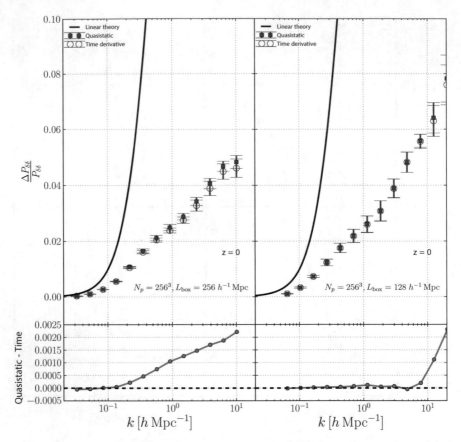

Fig. 5.6 (Left) $\Delta P/P$ for the lower-resolution (L_{256}) F6 simulations, shown at $z = 0$ in the top panel, with the relative difference between the quasi-static and non-static runs in the lower panel. There appears to be a systematic increase in the offset between the two cases with increasing k. We find that this discrepancy is purely a numerical artefact. (Right) The same is done for the L_{128} simulation for F6 at $z = 0$. It is clear to see that increasing the resolution has led to a much improved convergence between the quasi-static and non-static cases (except at the very largest k). Figure reproduced from Bose et al. (2015)

5.5.2 Configuration Space

So far we have focussed on the quantities describing the cosmic density and velocity fields in the Fourier space. Now, for completeness of our considerations, in this section we will focus on the configuration space. The clustering statistics of quantities defined in the configuration space provides a complementary picture of the field properties. The variance and the two-point correlation functions of a cosmic field are related to its Fourier power spectrum by:

Fig. 5.7 $\Delta P/P$ for the highest-resolution (L_{64}) F6 simulations, at $z = 0.5$. Note that this was simulated using only one realisation, so the error bars represent the scatter in each bin of log(k). One can already see the marked improvement in the agreement between the quasi-static and time derivative simulations, compared to equivalent redshift and even $z = 0$ for the L_{256} run (Fig. 5.3). Figure reproduced from Bose et al. (2015)

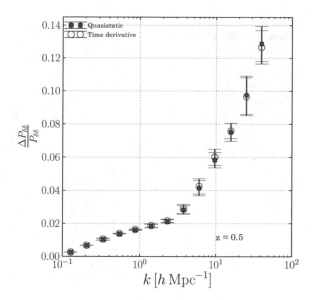

$$\xi(r) = \int \frac{dk}{2\pi^2} k^2 P(k) \frac{\sin(kr)}{(kr)}, \qquad (5.5.54)$$

$$\sigma^2(r) = \int \frac{dk}{2\pi^2} k^2 P(k) W_{\text{TH}}^2(kr). \qquad (5.5.55)$$

Here W_{TH} is the Fourier top-hat window and r is the comoving separation (or smoothing) scale in h^{-1} Mpc. We have computed both variance and two-point correlation function for the density and velocity divergence fields for all our L_{256} runs. For a set of smoothing scales satisfying $1 \leq r/(h^{-1}\,\text{Mpc}) \leq 0.1 L_{box}$ the denoted differences between quasi-static and time derivatives runs were even smaller then any of the differences we have observed for the density and velocity power spectra shown in figures from Figs. 5.1, 5.2, 5.3 and 5.4. Thus we can conclude that both frequency and configuration space two-point statistics used so far in this study are fostering a consistent picture. This reassures us that any differences in the properties of the density and velocity fields between quasi-static and time derivatives runs must be very small.

Hellwing et al. (2010); Hellwing et al. (2013) have indicated that the high-order moments are much more sensitive probes of even minute changes in the cosmic density field. They have shown in particular, that the clustering amplitudes are well posed to emphasise even very small differences in the clustering pattern when applied for modified gravity models. Following method of Hellwing et al. (2013), we have computed the reduced skewness $S_3^{\delta;\theta} \equiv \langle \delta^3; \theta^3 \rangle \sigma_{\delta;\theta}^{-4}$ and the reduced kurtosis $S_4^{\delta;\theta} \equiv \langle \delta^4; \theta^4 \rangle \sigma_{\delta;\theta}^{-6} - 3\sigma_{\delta;\theta}^{-2}$ for our ensemble of L_{256} simulations. For all the relevant smoothing scales we have not found any significant differences between quasi-static and time derivatives realisations in any of our runs.

Fig. 5.8 Probability density functions for the density field $\delta + 1$ in F6 at $z = 0.5$ computed within a spherical top-hat window, smoothed at $r = 0.25\ \mathrm{h}^{-1}$ Mpc. The distribution for the quasi-static simulation is shown in the blue line, whereas that of the non-static simulation is displayed in the red circles. Figure reproduced from Bose et al. (2015)

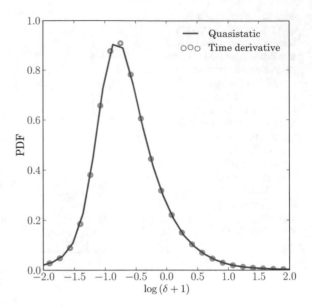

The results described in Sect. 5.5.1 augmented by our findings concerning the configuration space clustering statistics clearly demonstrate, that in the statistical sense the cosmic density and velocity fields produced in quasi-static and time derivative runs are equivalent down to resolved scales.

Finally, to summarise this section we show in Fig. 5.8 the probability distribution functions (PDFs) of the density field computed at $z = 0.5$ for our high-resolution L_{64} runs. Here we compare only the PDFs of the F6 brand modelled in our two approaches, with the smoothing scale, $r = 0.25\ \mathrm{h}^{-1}$ Mpc (equivalent to the size of one grid cell in L_{64}). Comparing the PDFs of the two realisations serves as our final test. So far we have focussed on statistical quantities, in which any signal coming from relatively small spatial regions would be strongly suppressed. One could imagine that there might exist some special regions in the density field, where the time derivatives of the scalaron could take bigger values and hence make a bigger impact the dynamics of the cosmic fields. The very centres of cosmic voids can serve as one example of such a place. The extremely low density in those locations could in principle allow for much stronger non-linear behaviour of the scalar field. However, analysis of the data plotted in the Fig. 5.8 evidently convinces us that both high and low δ tails of the PDF agree remarkably well in the compared simulations. All extreme objects, like very deep voids or very massive clusters, populate the aforementioned tails of the density PDF. The fact that the both curves agrees also in these regions guarantee that the scalaron and the matter fields exhibit the same dynamical evolution in both quasi-static and time derivative simulations.

5.6 Numerical Considerations

In this section, we discuss some of the code-specific numerical issues that one needs to account for in the interpretation of our results above.

5.6.1 Convergence of Solutions

In ECOSMOG, between successive relaxation sweeps, one can define a residual d^h, as the difference between the numerical values of the two sides of the equations being solved. Convergence (or alternatively, the signal to "stop" further relaxation iterations) is achieved when the residual gets smaller than some predefined threshold, the so-called *convergence criterion*. In practice, however, the accuracy one would ever achieve when numerically solving our partial differential equation is fundamentally limited by a numerical error, the so-called truncation error τ^h, imposed by the discretisation of the continuous differential equation. The latter implies that there is no point to further reduce the residual by doing more relaxation iterations, once it has become smaller than the truncation error (Press et al. 2002):

$$\left| d^h \right| \leq \alpha \left| \tau^h \right| , \qquad (5.6.56)$$

where α is some constant ($\sim 1/3$).

Throughout this work, convergence is deemed to have been achieved when the residual $|d^h| \leq 10^{-8}$, which is a significantly stronger criterion than that in Eq. 5.6.56, and further reducing $|d^h|$ does not change the results by much. If, however, one uses $\left| d^h \right| \leq \alpha \left| \tau^h \right|$, then the results will be changed, and the change itself is larger than the offsets caused by including the time derivatives. Obviously, this is a change that we have no control over. The quasi-static approximation therefore introduces an error well below that caused by the discretisation of the differential equation itself.[2]

5.6.2 Box Size and Resolution

As we have seen in the previous section, the results that we get for the F6 simulations depend on the resolution. This is quite an odd result, and on first instance, slightly contrary to what one might expect when considering the CFL condition in Eq. 5.4.45,

---

[2]It is often argued that one should make $|d^h|$ as small as practically possible, instead of stopping at $|d^h| \sim \tau^h/3$, to prevent the numerical errors in solving the differential equation at individual steps from accumulating over the many time steps of a simulation. While this is true to a certain extent, it is not clear that the discretisation error itself will not accumulate in this case (recall that, if d^h could be brought to zero, then the remaining error is completely from the discretisation). Again, the way to get away from this problem is to reduce the discretisation error by increasing the (spatial) resolution, and then check for convergence.

which is what ECOSMOG uses to determine the size of the integration time step. Reducing the box size (as we have done here) will reduce dx by the same factor, but the integration time step is also affected in the same way to ensure that particles do not move more than a cell size during one time step. As such, one would expect that adjusting the resolution by means of a increase or decrease in the size of the simulation box should not affect how different the time derivative case is from the quasi-static limit. Why then do the higher-resolution L_{128} (Fig. 5.5) and L_{64} runs reduce the discrepancy of this offset seen in the L_{256} run (Fig. 5.3)?

As a result of decreasing the time step in the higher resolution simulations, the particles do not travel as far as they do in low-resolution simulations in a given time step, and so between two consecutive steps, the f_R field configuration in real space does not change as much as in a full time step run, which makes its time derivatives smaller. In terms of the equations of motion (say in Eq. 5.4.38), this amounts to saying that the value $u_{i,j,k}^{(n)} - u_{i,j,k}^{(n-1)}$ does not change as much when the time step is reduced. We have tested this in the $L_{256/2}$ run, by re-running the L_{256} simulation in F6, this time artificially halving the time step that ECOSMOG would naturally use (keeping the force resolution the same), and find that the offset between the quasi-static and time derivative runs is indeed reduced as in the L_{128} simulations.

We thus conclude that with increased resolution, the reduced time steps make the quasi-static and non-static F6 simulations converge better, and in the convergence limit the time derivatives do not have a big impact on any of our $f(R)$ gravity simulations.

5.6.3 Inaccuracies Due to Averaging over Oscillations

One of the major caveats behind our analysis is the manner in which we include the time evolution of f_R in our simulations. Since $m_{\text{eff}} \gg H$, the scalaron field is expected to oscillate very fast about its minimum as it evolves. As mentioned in Sect. 5.4.2, the time derivative is calculated by averaging f_R over the many oscillations. We can make a crude estimate of the error caused by this procedure by following the methodology of Brax et al. (2012b).

The background evolution of the field scalaron is given by the equation:

$$\ddot{f}_R + 3H\dot{f}_R + \frac{\text{d}V_{\text{eff}}}{\text{d}f_R} = 0 , \qquad (5.6.57)$$

where $\text{d}V_{\text{eff}}/\text{d}f_R = -1/3\,(R - f_R R + 2f(R) + 8\pi G\rho_m)$ as defined in Eq. 5.2.7. Now, let us consider small perturbations of the scalaron about its minimum as $\delta f_R \equiv f_R - f_{R,\text{min}}$ (note that across this subsection δf_R is *not* the spatial perturbation), and derive the following evolution equation for δf_R:

$$\delta \ddot{f}_R + 3H\delta \dot{f}_R + m_{\text{eff}}^2 \delta f_R = F(t)$$
$$\equiv -\frac{1}{a^3}\frac{\mathrm{d}}{\mathrm{d}t}\left[a^3 \frac{\mathrm{d}f_{R,\text{min}}}{\mathrm{d}t}\right]. \qquad (5.6.58)$$

The minimum equation for $f(R)$ gravity is given by:

$$\left.\frac{\mathrm{d}V_{\text{eff}}}{\mathrm{d}f_R}\right|_{f_{R,\text{min}}} = 0\,, \qquad (5.6.59)$$

which has been used to derive the above equation and which also implies that (by taking the time derivative of the relation $R \approx -8\pi G\rho_m$):

$$\frac{\mathrm{d}f_{R,\text{min}}}{\mathrm{d}t} \approx -\frac{8\pi G\rho_m}{m_{\text{eff}}^2}H\,. \qquad (5.6.60)$$

The time-dependent force term $F(t)$ then becomes:

$$F(t) \approx \frac{8\pi G\rho_{m0}}{a^3}\frac{\mathrm{d}}{\mathrm{d}t}\left[\frac{H}{m_{\text{eff}}^2}\right]. \qquad (5.6.61)$$

In addition to being driven by the (slow) time evolution of the minimum $f_{R,\text{min}}$, the scalaron field also experiences a number of "kicks" when relativistic species become non-relativistic and thus starts to contribute to $T_\mu^\mu \approx \rho_m$. Because the transition from relativistic to non-relativistic happens on a relatively short time scale compared to the Hubble time, we can model this effect as "instantaneous kicks" (Brax et al. 2004)[3]:

$$F \approx \frac{8\pi G\rho_{m0}}{a^3}\frac{\mathrm{d}}{\mathrm{d}t}\left[\frac{H}{m_{\text{eff}}^2}\right] - \beta \sum_j \kappa_j H_j \delta(t - t_j), \qquad (5.6.62)$$

where t_j is the time at which the transition from relativistic to non-relativistic happens and $\kappa_j \approx g/g_\star(m_j) \leq 1$, with g the number of degrees of freedom of the species that is becoming non-relativistic and $g_\star(m_j)$ the number of relativistic species at time t_j, when the temperature T is equal to the mass m_j. $\beta \sim \mathcal{O}(1)$ is a constant and H_j is the Hubble expansion rate at t_j.

In what follows, we limit ourselves to the time of the electron decoupling, t_e, as an example of the analysis:

$$\delta \ddot{f}_R + 3H\delta \dot{f}_R + m_{\text{eff}}^2 \delta f_R \approx \frac{8\pi G\rho_{m0}}{a^3}\frac{\mathrm{d}}{\mathrm{d}t}\left[\frac{H}{m_{\text{eff}}^2}\right]$$
$$- \beta_e \kappa_e H_e \delta(t - t_e)\,. \qquad (5.6.63)$$

---

[3]The kick is by the sudden increase in the non-relativistic ρ_m, as can be seen from $\mathrm{d}V_{\text{eff}}/\mathrm{d}f_R = -[R - f_R R + 2f(R) + 8\pi G\rho_m]/3$ – because of the quick change in ρ_m, $f_{R,\text{min}}$ is changed while the true f_R needs time to respond to this.

Defining a new field ψ which satisfies $\delta f_R = a^{-3/2}\psi$, this equation can be rewritten as:

$$\ddot{\psi} + \left[m_{\text{eff}}^2 + \frac{9}{4}wH^2 \right]\psi \approx 8\pi G\rho_{m0}a^{-3/2}\frac{\mathrm{d}}{\mathrm{d}t}\left[\frac{H}{m_{\text{eff}}^2} \right]$$
$$- \beta_e\kappa_e H_e a_e^{3/2}\delta\left(t - t_e \right), \tag{5.6.64}$$

where $w = P/\rho$ is the effective equation of state, with ρ, P including contributions from all matter species. Since $m^2 \gg H^2$, we can solve Eq. 5.6.64 using the Wentzel–Kramers–Brillouin (WKB) approximation, and finally get:

$$\delta f_R \approx \frac{8\pi G\rho_{m0}}{m_{\text{eff}}^2 a^3}\frac{\mathrm{d}}{\mathrm{d}t}\left[\frac{H}{m_{\text{eff}}^2} \right] \tag{5.6.65}$$
$$-\Theta\left(t - t_e \right)\beta_e\kappa_e H_e \frac{a_e^{3/2}}{a^{3/2}}\frac{1}{\sqrt{m_e m_{\text{eff}}}}\sin\int_{t_e}^{t} m\left(t' \right)\mathrm{d}t' \ .$$

where $\Theta\left(t - t_e \right)$ is the Heaviside function, $m_e \equiv m_{\text{eff}}(t = t_e)$, $H_e = H(t = t_e)$ and similarly for β_e and κ_e.

By rewriting:

$$\frac{\mathrm{d}}{\mathrm{d}t}\left[\frac{H}{m_{\text{eff}}^2} \right] = g(t)\frac{H^2}{m_{\text{eff}}^2}\ , \tag{5.6.66}$$

and using:

$$8\pi G\rho_{m0} = 3H_0^2\Omega_m, \tag{5.6.67}$$

we can finally average over the rapid oscillations to get:

$$\langle \delta f_R^2 \rangle(t) \approx \frac{9\Omega_m^2 g^2(t)}{a^6}\frac{H_0^4}{m_{\text{eff},0}^4}\frac{m_{\text{eff},0}^4}{m_{\text{eff}}^4(t)}\frac{H^4}{m_{\text{eff}}^4(t)}$$
$$+ \frac{\beta_e^2\kappa_e^2}{2}\frac{a_e^3}{a^3}\frac{H_e^2}{m_e^2}\frac{m_e}{m_{\text{eff}}(t)}, \tag{5.6.68}$$

where, again, a subscript $_0$ denotes the value at present day.

At late times, e.g., $a \sim 1$, the first term in the above expression is of order $(H_0/m_{\text{eff},0})^8$ and is extremely small (compared to $|f_{R0}|$) because $H_0/m_{\text{eff},0}$ is typically less than 10^{-3} for the models studied here. This term appears because of the shift of $f_{R,\text{min}}$, which itself is due to the evolution of the background matter density in the Universe. It has nothing to do with the oscillations that we are interested in here.

The second term characterises the amplitude of the oscillations of δf_R. Up until the onset of the acceleration phase, we have $|\bar{f}(R)| \ll 8\pi G\bar{\rho}_m$ and therefore

$\bar{R} \approx -8\pi G \bar{\rho}_m$, where $\bar{\rho}_m$ has no contribution from radiation even in the radiation-dominated era. This relation $m_{\text{eff}}^2(t) \approx -(1/3)\mathrm{d}R/\mathrm{d}f_R$ gives:

$$m_{\text{eff}}^2(t) \approx \frac{H_0^2 \Omega_m}{3n(n+1)\xi} \left[\frac{-R}{M^2} \right]^{n+2}$$

$$\approx 3^{n+1} \frac{H_0^2 \Omega_m}{n(n+1)\xi} a^{-3(n+2)}. \tag{5.6.69}$$

By noting that $n = 1, 3^{n+1}\Omega_m \sim \beta_e^2 \kappa_e^2 \sim \mathscr{O}(1)$, we can combine the above two equations to estimate the amplitude of the oscillation as:

$$\langle \delta f_R^2 \rangle^{1/2}(t) \sim \xi a_e^{7/2} \Omega_r a^{3/2} \tag{5.6.70}$$

where $\xi \approx 34^2 |\bar{f}_{R0}|$, $\Omega_r \sim 10^{-4}$ is the present-day fractional energy density of radiation and $a_e \sim 10^{-9}$ is the scale factor at t_e. The late-time dominance of dark energy slightly alters the relation $\bar{R} \approx -8\pi G \bar{\rho}_m$, but nevertheless the above result still serves as a good order-of-magnitude estimate.

We are more interested in the quantity:

$$\frac{\langle \delta f_R^2 \rangle^{1/2}(t)}{|\bar{f}_{R,\min}(t)|} \sim 9 a_e^{7/2} \Omega_r a^{-9/2}, \tag{5.6.71}$$

which is independent of $|\bar{f}_{R,0}|$ and decays over time. A quick calculation shows that for our simulations ($z < 49$) the amplitude of the oscillation is always smaller than 10^{-27} times $\bar{f}_{R,\min}$, with a value of 10^{-35} today.[4]

Evidently, with such tiny amplitudes, the oscillations are unlikely to have any impact on our result, and the averaging over many oscillations should work accurately. Note also that the smallness of $\langle \delta f_R^2 \rangle^{1/2}(t)$ implies that it is probably unrealistic to follow the oscillations using explicit time integration in a numerical simulation poised for the study of cosmic structure formation, such as ours here.

Of course, the analysis in this subsection has been greatly simplified. In reality, the situation could be much more complicated. For example, the scalaron field f_R at a given position of space may not be oscillating around the minimum of its effective potential as determined by matter density at that position, but instead far from that minimum due to interactions with the density field in the environment; the oscillations could have a position (or local-density) dependent mass $m_{\text{eff}}(t, \mathbf{x})$; and there can even be 'micro kicks' caused by rapid changes of local matter density due to particles moving to or away from the position, etc.. Such 'micro kicks' may not be well approximated as instantaneous kicks because particle velocity $v \ll c$, and they have already been accounted for in our time integration scheme.

---

[4]Note that we can use a_e in the above expressions and estimates, because electrons are the last species of standard-model particles that become non-relativistic.

5.6.4   Initial Conditions

We see from Figs. 5.1, 5.2, 5.3 and 5.5 that the different initial conditions can lead to significant variations in the results. This can be seen in the form of error bars on the data points in the figures, which represent the scatter within each k-bin over the five realisations – the relative enhancement of the power spectra $\Delta P/P$ can be lower or higher than the mean of the bin. Our results demonstrate that the variations across different realisations dominate the differences induced by including time derivatives.

5.6.5   The Effect of Baryons

In this chapter, we have ignored the effect of baryons in our simulations. While this is not expected to make much of a difference on large scales, the baryonic effects are more pronounced on non-linear scales, making it more difficult to correctly measure the power spectrum $P(k)$ in this regime. van Daalen et al. (2014) found that there can be a discrepancy of more than 10% in the two-point correlation function on sub-Mpc scales between dark matter only simulations, and those with baryonic effects included. The difference between the inclusion and non-inclusion of time derivatives in our $f(R)$ gravity simulations is typically sub-percent, so we expect that any errors from the non-inclusion of baryons significantly dominate those caused by the quasi-static approximation.

5.7   Summary and Discussions

In this chapter, we have studied the effect of including time derivatives in the scalar field equation of motion in numerical simulations of structure formation for $f(R)$ gravity, which is a departure from the quasi-static approximation usually used in such simulations. To this end, we have generalised both the $f(R)$ equation itself (Sect. 5.3.2) and the Poisson equation (Sect. 5.3.3), which are the equations that govern the formation of cosmic structures in this model. We find that, in both cases, the inclusion of time derivatives results in additional terms entering the equations compared to the quasi-static case, as seen in Eq. 5.4.35. To solve these equations, we make use of ECOSMOG, using 256^3 particles in different boxes (of size $256\,h^{-1}$ Mpc, $128\,h^{-1}$ Mpc and $64\,h^{-1}$ Mpc), to test for the effects of resolution. In the low-resolution case, we evolve three different Hu–Sawicki $f(R)$ models: F4, F5 and F6, corresponding to different values of the scalaron field $|\bar{f}_{R0}|$ (Sect. 5.5).

By looking at the enhancement of the matter and velocity divergence power spectra relative to ΛCDM, we find that, in the cases of F4 and F5 (Sect. 5.5.1.1), the low-resolution L_{256} box simulations confirm that including time derivatives introduces only an insignificant difference from the quasi-static approximation, whereas this difference is larger in the case of F6. To see if changing box size has any effect on this discrepancy, we perform the F6 simulations in the L_{128} runs (Sect. 5.5.1.2), and find that this large offset becomes smaller. To verify whether this is actually a consequence of increasing the resolution, we also run two additional tests. The first is a variation of the original L_{256} simulation but with its time steps artificially halved (which we dub the $L_{256/2}$ simulation). This simulation has the same mass and force resolution as the low-resolution L_{256} runs, but it shows the same decrease of the non-static effect as in L_{128}. The second is an ever higher resolution simulation with 256^3 particles in a box of size $64\,h^{-1}$Mpc, which we call L_{64}. This simulation has even smaller time steps and shows even better agreement between the quasi-static and non-static runs. Finally, we test the statistics of the configuration space for both the static and non-static cases, and again find no discernible differences.

The implications of the additional tests are twofold:

- They confirm that with increasing temporal resolution, our implicit scheme for time integration does converge, and this is a nontrivial check that our new code and algorithm works consistently;
- The converged result is that, even for F6, the inclusion of time derivative is neither crucial nor necessary, and that the quasi-static approximation works reasonably well for all $f(R)$ models studied here.

We have also discussed numerical issues associated with our algorithm. In particular, our time-integration scheme assumes implicitly that the code actually evolves quantities which are averaged over many scalaron field oscillations. Our qualitative analysis shows that the amplitudes of such oscillations, although grow in time, are much smaller than the average value (i.e., the oscillation centre) at all epochs of interest to us, and as a result the implicit time-average should have no impact on our result in practice. We have also discussed other intrinsic sources of scatter, such as the different initial conditions (cosmic variance) and the convergence criterion for our relaxation method, and concluded that they are all significantly larger than the possible error caused by the quasi-static approximation.

To summarise: we find that the effects of the scalar field time derivatives are so small that can be safely neglected for the most practical applications in cosmology.

The three models we consider – F4, F5 and F6 – span a wide range in the strength of the screening mechanism, from very weak to very strong, but in all these cases the quasi-static approximation holds yielding reliable results. In particular, F4 corresponds to a model where the chameleon screening is so weak that it is closer to unscreened theories such as coupled quintessence (Li and Barrow 2011a, b), and the conclusion can be generalised to those classes of theories.

On the other hand, we must be cautious when trying to generalise the conclusion here to other modified gravity theories. An important example is the Galileon gravity model (Nicolis et al. 2009; Deffayet et al. 2009), which has the Dvali–Gabdadze–Porrati (DGP) model (Dvali et al. 2000) as a subclass. Barreira et al. (2013); Li et al. (2013) found that neglecting the time derivatives results in the equations having no real solutions in low-density regions, which does not occur in the case of $f(R)$ gravity. As a result, for those theories, the time derivatives are likely to have a non-negligible impact on the cosmic fields. It would be interesting to apply our method of including non-static effects to Galileon simulations and quantify this impact, and this will be left for future work.

5.8 New Developments Since Submission of This Thesis in April 2017

In this chapter, we have focussed on relaxing the quasi-static approximation in $f(R)$ gravity theories, where we found that the approximation, even in the highly non-linear regime is in fact a very good one.

A similar exercise to the one described in this thesis was performed by Llinares and Mota (2014b) in the case of the symmetron model. In this scalar field theory, the action is given by:

$$ S = \int \sqrt{-g} \left[\frac{M_{Pl}^2}{2} R - \frac{1}{2} \nabla^a \phi \nabla_a \phi - V(\phi) \right] d^4x + \int \mathscr{L}_m(\tilde{g}_{\mu\nu}) d^4x, \tag{5.8.72} $$

where we have defined the symmetron field, ϕ, and $\tilde{g}_{\mu\nu} = A^2(\phi) g_{\mu\nu}$, where $A(\phi)$ is composed of characteristic scales in this model. In this model, the scalar field equation of motion becomes:

$$ \ddot{\phi} + 3H\dot{\phi} - \frac{1}{a^2}\nabla^2\phi = -V_{,\phi} - A_{,\phi}\rho_m. \tag{5.8.73} $$

In their analysis, Llinares and Mota (2014b) find that while the globally-averaged matter power spectrum shows little difference when comparing the static and non-static solutions to the symmetron equations (typically of the order of $\sim 0.2\%$, the power spectrum measured *locally* (i.e., filtered on different smoothing scales) can show substantial differences that rise to $\sim 1\%$. These differences are attributed to the formation of domain walls in the symmetron model, resulting in different phenomenology depending on whether a particular region of space is located inside or outside the domain walls. In more recent work by Hagala et al. (2017), it was shown that non-static effects in the symmetron model can result in the generation of 'waves' sourced by the scalar field, thereby unscreening regions that would have otherwise been screened in the static case. Unlike $f(R)$ gravity, therefore, incorporating non-static effects is crucial for self-consistently evolving the dynamics of the symmetron field.

References

Barreira A, Li B, Hellwing WA, Baugh CM, Pascoli S (2013) JCAP 10:027. https://doi.org/10.1088/1475-7516/2013/10/027, http://adsabs.harvard.edu/abs/2013JCAP...10.027B

Bose S, Hellwing WA, Li B (2015) JCAP 2:034. https://doi.org/10.1088/1475-7516/2015/02/034, http://adsabs.harvard.edu/abs/2015JCAP...02.034B

Brax P, van de Bruck C, Davis A-C, Khoury J, Weltman A (2004) Phys D Rev 70:123518. https://doi.org/10.1103/PhysRevD.70.123518, http://adsabs.harvard.edu/abs/2004PhRvD.70l3518B

Brax P, van de Bruck C, Davis A-C, Shaw D (2010) Phys Rev D 82:063519. https://doi.org/10.1103/PhysRevD.82.063519

Brax P, Davis A-C, Li B, Winther HA (2012) Phys D Rev 86:044015. https://doi.org/10.1103/PhysRevD.86.044015, http://adsabs.harvard.edu/abs/2012PhRvD.86d4015B

Cai Y-C, Li B, Cole S, Frenk CS, Neyrinck M (2014) MNRAS 439:2978. https://doi.org/10.1093/mnras/stu154, http://adsabs.harvard.edu/abs/2014MNRAS.439.2978C

Carroll SM, de Felice A, Duvvuri V, Easson DA, Trodden M, Turner MS (2005) Phys D Rev 71:063513. https://doi.org/10.1103/PhysRevD.71.063513, http://adsabs.harvard.edu/abs/2005PhRvD.71f3513C

Cautun MC, van de Weygaert R (2011) The DTFE public software: the delaunay tessellation field estimator code, Astrophysics source code library. arXiv:1105.0370

Clifton T, Ferreira PG, Padilla A, Skordis C (2012) Phys Rep 513:1. https://doi.org/10.1016/j.physrep.2012.01.001, http://adsabs.harvard.edu/abs/2012PhR...513....1C

Colombi S, Jaffe A, Novikov D, Pichon C (2009) MNRAS 393:511. https://doi.org/10.1111/j.1365-2966.2008.14176.x, http://adsabs.harvard.edu/abs/2009MNRAS.393.511C

Courant R, Friedrichs K, Lewy H (1928) Math Ann 100:32. https://doi.org/10.1007/BF01448839, http://adsabs.harvard.edu/abs/1928MatAn.100...32C

Deffayet C, Esposito-Farèse G, Vikman A (2009) Phys D Rev 79:084003. https://doi.org/10.1103/PhysRevD.79.084003, http://adsabs.harvard.edu/abs/2009PhRvD.79h4003D

Dvali G, Gabadadze G, Porrati M (2000) Phys Lett B 485:208. https://doi.org/10.1016/S0370-2693(00)00669-9, http://adsabs.harvard.edu/abs/2000PhLB..485..208D

Hagala R, Llinares C, Mota DF (2017) Phys Rev Lett 118:101301. https://doi.org/10.1103/PhysRevLett.118.101301, http://adsabs.harvard.edu/abs/2017PhRvL.118j1301H

He J-H, Li B, Jing YP (2013) Phys D Rev 88:103507. https://doi.org/10.1103/PhysRevD.88.103507, http://adsabs.harvard.edu/abs/2013PhRvD.88j3507H

Hellwing WA, Juszkiewicz R, van de Weygaert R (2010) Phys D Rev 82:103536. https://doi.org/10.1103/PhysRevD.82.103536, http://adsabs.harvard.edu/abs/2010PhRvD.82j3536H

Hellwing WA, Li B, Frenk CS, Cole S (2013) MNRAS 435:2806. https://doi.org/10.1093/mnras/stt1430, http://adsabs.harvard.edu/abs/2013MNRAS.435.2806H

Hellwing WA, Barreira A, Frenk CS, Li B, Cole S (2014) Phys Rev Lett 112:221102. https://doi.org/10.1103/PhysRevLett.112.221102, http://adsabs.harvard.edu/abs/2014PhRvL.112v1102H

Hinshaw G et al (2013) APJS 208:19. https://doi.org/10.1088/0067-0049/208/2/19, http://adsabs.harvard.edu/abs/2013ApJS.208...19H

Hinterbichler K, Khoury J (2010) Phys Rev Lett 104:231301. https://doi.org/10.1103/PhysRevLett.104.231301, http://adsabs.harvard.edu/abs/2010PhRvL.104w1301H

Hu W, Sawicki I (2007) Phys D Rev 76:064004. https://doi.org/10.1103/PhysRevD.76.064004, http://adsabs.harvard.edu/abs/2007PhRvD.76f4004H

Jennings E, Baugh CM, Pascoli S (2011) MNRAS 410:2081. https://doi.org/10.1111/j.1365-2966.2010.17581.x, http://adsabs.harvard.edu/abs/2011MNRAS.410.2081J

Khoury J (2010). arXiv:1011.5909

Li B, Barrow JD (2011a) Phys D Rev 83:024007. https://doi.org/10.1103/PhysRevD.83.024007, http://adsabs.harvard.edu/abs/2011PhRvD.83b4007L

Li B, Barrow JD (2011b) MNRAS 413:262. https://doi.org/10.1111/j.1365-2966.2010.18130.x, http://adsabs.harvard.edu/abs/2011MNRAS.413.262L

Li B, Zhao G-B, Teyssier R, Koyama K (2012) JCAP 1:051. https://doi.org/10.1088/1475-7516/2012/01/051, http://adsabs.harvard.edu/abs/2012JCAP...01.051L

Li B, Barreira A, Baugh CM, Hellwing WA, Koyama K, Pascoli S, Zhao G-B (2013a) JCAP 11:012. https://doi.org/10.1088/1475-7516/2013/11/012, http://adsabs.harvard.edu/abs/2013JCAP...11.012L

Li B, Hellwing WA, Koyama K, Zhao G-B, Jennings E, Baugh CM (2013b) MNRAS 428:743. https://doi.org/10.1093/mnras/sts072, http://adsabs.harvard.edu/abs/2013MNRAS.428.743L

Llinares C, Mota DF (2014a) Phys D Rev 89:084023. https://doi.org/10.1103/PhysRevD.89.084023, http://adsabs.harvard.edu/abs/2014PhRvD.89h4023L

Llinares C, Mota DF (2014b) Phys D Rev 89:084023. https://doi.org/10.1103/PhysRevD.89.084023, http://adsabs.harvard.edu/abs/2014PhRvD.89h4023L

Mota DF, Shaw DJ (2007) Phys D Rev 75:063501. https://doi.org/10.1103/PhysRevD.75.063501, http://adsabs.harvard.edu/abs/2007PhRvD.75f3501M

Nicolis A, Rattazzi R, Trincherini E (2009) Phys D Rev 79:064036. https://doi.org/10.1103/PhysRevD.79.064036, http://adsabs.harvard.edu/abs/2009PhRvD.79f4036N

Perlmutter S et al (1999) ApJ 517:565. https://doi.org/10.1086/307221, http://adsabs.harvard.edu/abs/1999ApJ...517.565P

Press WH, Teukolsky SA, Vetterling WT, Flannery BP (2002) Numerical recipes in C++ : the art of scientific computing

Riess AG et al (1998) AJ 116: 1009. https://doi.org/10.1086/300499, http://adsabs.harvard.edu/abs/1998AJ....116.1009R

Schaap WE, van de Weygaert R (2000) A&A 393:L29. http://adsabs.harvard.edu/abs/2000A%26A...363L..29S

Schmidt F, Vikhlinin A, Hu W (2009b) Phys D Rev 80:083505. https://doi.org/10.1103/PhysRevD.80.083505, http://adsabs.harvard.edu/abs/2009PhRvD.80h3505S

Teyssier R (2002) A&A 385:337. https://doi.org/10.1051/0004-6361:20011817, http://adsabs.harvard.edu/abs/2002A%26A...385.337T

van Daalen MP, Schaye J, McCarthy IG, Booth CM, Dalla Vecchia C (2014) MNRAS 440:2997. https://doi.org/10.1093/mnras/stu482, http://adsabs.harvard.edu/abs/2014MNRAS.440.2997V

Chapter 6
Speeding up N-Body Simulations of Modified Gravity: Chameleon Screening Models

6.1 Introduction

[1]Modified gravity theories (Clifton et al. 2012; Joyce et al. 2016) are popular alternatives to the cosmological constant and dark energy models (Copeland et al. 2006) to explain the observed accelerating expansion of our Universe (Guy et al. 2010; Percival et al. 2010; Beutler et al. 2011; Reid et al. 2012; Hinshaw et al. 2013; Riess et al. 2009). Rather than invoking a cosmological constant (Λ),[2] or a new energy component to drive the dynamics of the cosmos, these theories suggest that the Universe contains only normal and dark matter (which is often assumed as cold dark matter, or CDM), but the law of gravitation deviates from that prescribed by Einstein's General Relativity (GR) on large scales, resulting in an acceleration of the expansion rate.

Since the law of gravity is universal, deviations from GR on large scales are often associated with changes in the behaviour on small scales. Any such small scale changes, however, must be vanishingly small due to the strong constraints placed by numerous local tests of gravity (Will 2014). Consequently, viable modified gravity theories usually have some mechanism by which such modifications are suppressed, recovering GR in dense regions like the Solar System, where those gravity tests have been carried out and their resulting constraints apply. These are commonly referred to as 'screening mechanisms' in the literature, and are an inherent (instead of an add-on) property which comes from the dynamics of the theory. The screening effect implies that gravity behaves differently in different environments; this environmental dependence is often reflected in strong non-linearities in the field equations, which make both analytical and numerical studies of such theories challenging.

---

[1]The content of this Chapter is based on the article Bose et al. 'Speeding up *N*-body simulations of modified gravity: chameleon screening models', Journal of Cosmology and Astroparticle Physics, Volume 2015, Issue 2, published 24 February 2015. Reproduced with permission. All rights reserved, https://doi.org/10.1088/1475-7516/2017/02/050.

[2]Note, however, that in many modified gravity theories, such as the one studied in this thesis, an effective cosmological constant is still required to drive the accelerated expansion.

© Springer International Publishing AG, part of Springer Nature 2018
S. Bose, *Beyond ΛCDM* , Springer Theses,
https://doi.org/10.1007/978-3-319-96761-5_6

In most theories that are currently being investigated, the modification to GR boils down to an extra (so-called fifth) force that is mediated by a new scalar field, and screening in this context means suppression of the fifth force. In one class of such theories, this is achieved by a coupling of the scalar field to matter and a non-linear self-interaction potential of the scalar field. With appropriate choices of the coupling and potential, the dynamics of the scalar field can ensure that, in high density regions, the fifth force it mediates decays exponentially fast with distance, or becomes extremely small in its amplitude. Chameleon theories (Weltman 2004; Mota and Shaw 2007), with $f(R)$ gravity (Faraoni 2010) (see also Barrow 2007, Hu and Sawicki 2007, Brax et al. 2008) as a representative example, is an instance of the former case, while the dilaton (Brax et al. 2010) and symmetron (Hinterbichler and Khoury 2010) models belong to the latter case.

Amongst the chameleon models, $f(R)$ gravity is currently the most well-studied case, and there exist numerous works investigating in detail its predictions for large-scale structure formation in the non-linear regime. This has been made possible by the continuous development of N-body simulation codes (e.g., Oyaizu 2008, Oyaizu et al. 2008, Schmidt et al. 2009a, Zhao 2009, 2010, Zhao et al. 2010, Zhao et al. 2011a, b, Hu 2011, Li et al. 2012b; Li et al. 2013b, Lombriser et al. 2012, Lee et al. 2013, Jennings et al. 2012). An efficient code amongst these is ECOSMOG (Li et al. 2012), based on the publicly available N-body and hydro code RAMSES (Teyssier 2002), which makes large simulations for $f(R)$ gravity feasible. Using the generic parameterisation for modified gravity theories (Brax et al. 2012; Brax et al. 2012c), ECOSMOG was extended to incorporate chameleon, dilaton and symmetron models (Brax et al. 2012a, 2013) in general. ECOSMOG has recently been compared with other codes developed subsequently, including MG-GADGET (Puchwein et al. 2013), Isis (Llinares et al. 2014) and MG- ENZO (Wilcox et al. 2016) and very good agreement was found between all these codes (Winther et al. 2015).

There are other modified gravity theories, such as the Dvali–Gabadadze–Porrati (Dvali et al. 2000) (DSP) brane-world model, in which screening is achieved by non-linear derivative self-couplings of a scalar field. Well-studied examples include the K-mouflage (Brax and Valageas 2014a, b) and Vainshtein (1972) mechanisms, the latter being originally studied in massive gravity theories as a means to suppress the extra helicity modes of massive gravitons so that GR is recovered in the massless limit. In addition to the non-linear massive gravity (de Rham et al. 2011; Sbisà et al. 2012; Chkareuli and Pirtskhalava 2012) and braneworld models, the Vainshtein mechanism is also employed in general setups, such as the Galileon models (Nicolis et al. 2009; Deffayet et al. 2009), which have been the subject of various recent studies (e.g. Chow et al. 2009, Silva and Koyama 2009, Ali et al. 2010, Brax et al. 2011, Barreira et al. 2012, Falck et al. 2015, Barreira et al. 2017, Neveu et al. 2017).

The first two generations of modified gravity simulation codes (e.g., Oyaizu 2008, Llinares et al. 2008, Zhao 2010, Zhao et al. 2011a) were either not parallelised or had a uniform resolution across the whole simulation box, resulting in insufficient resolution and inefficiency. The current generation of codes, such as ECOSMOG, MG-GADGET, Isis and MG- ENZO, are all efficiently parallelised. These codes solve the non-linear field equations in modified gravity on meshes (or their equivalents), and employ the adaptive mesh refinement (AMR) technique to generate ever finer meshes in high density regions to increase resolution. However, even with these parallelised codes, modified gravity simulations currently are still very slow compared to the fiducial GR case. As we shall discuss below, this is partially due to the non-linear nature of the equations to be solved, and partly due to the specific numerical algorithms used. The greater computational cost of modified gravity simulations makes it difficult to achieve the resolution and volume attained in state-of-the-art simulations of standard gravity.

The coming decade will see a flood of high-precision observational data from a new generation of cosmological surveys, such as eROSITA (Merloni et al. 2012), the Dark Energy Spectroscopic Instrument (DESI) (Levi 2013; DESI Collaboration 2016a,b), EUCLID (Laureijs 2011) and the Large Synoptic Survey Telescope (LSST) (Ivezic 2008). These surveys will provide us with golden opportunities to perform cosmological tests of gravity (see Ref. Koyama 2016 for a recent review) and seek a better understanding of the origin of cosmic acceleration. As things stand now, it is the lack of more powerful simulation methods that limits the accuracy and size that modified gravity simulations can possibly attain, therefore preventing us from fully exploiting future observations. This has led to many attempts to speed up simulations using approximate methods (e.g., Winther and Ferreira 2015, Barreira et al. 2015), or develop alternative methods to predict theoretical quantities (e.g., Li and Efstathiou 2012, Zhao 2014, Mead et al. 2015, Cataneo et al. 2016). These alternative methods are fast substitutes of full simulations and powerful when quickly exploring a large parameter space is the primary concern. However, simulations are nevertheless necessary to calibrate these methods or when better (e.g., %-level) accuracy is needed, as well as to study the impact of different theories of gravity on galaxy formation.

In Barreira et al. (2015), an approximate method to speed up N-body simulations of Vainshtein-type models was presented and shown to reduce the overhead[3] of solving the modified gravity equation to the level of $50 \sim 100\%$, with the errors in various cosmological quantities being controlled to well under $\sim 1\%$ or smaller (comparable to the discrepancies in the predictions of different modified gravity simulation codes Winther et al. 2015). The same method, however, does not work as accurately in chameleon-type models (see Appendix A.1), the simulations for which are much more expensive than those for the Vainshtein-type models. Given that chameleon models are a large class of modified gravity models that are of interest to the theoretical and observational community, there is an equally urgent need for fast simulation methods for them – this is precisely the purpose of this work.

Unlike the truncated simulation method in Barreira et al. (2015), which artificially suppresses the solver of the modified gravity equation on higher refinement levels of the AMR meshes, and instead interpolates the solution on lower (or coarser)

---

[3]Throughout this chapter, the term 'overhead' is used to refer to the *extra* computational time (using the same machine and number of cores) involved in running a modified gravity simulation compared to standard gravity. For example, an overhead of 110% means that the modified gravity run requires $2.1 \times$ the CPU time of a ΛCDM simulation.

refinement levels to find *approximate* solutions on higher levels, the method proposed here still solves the *full* modified gravity equations on *all* levels. The improved efficiency comes instead from a different way to discretise the equation on meshes, that makes it less non-linear and greatly enhances the rate of convergence of the solution. The new scheme boosts the performance of the code *by a factor of 5* for a simulation with a periodic box of size 512 Mpc/h and 512^3 particles, and by a factor of *more than 20*, for a higher resolution setup with a box size of 128 Mpc/h and 512^3 particles. The method has its own limitation, namely that the existence of analytical solutions is a particular property of Hu–Sawicki (HS) $f(R)$ gravity – as well as a few other examples of chameleon, symmetron (see Appendix A.2) and dilaton – models. However, the generic nature of the HS model (in the sense that with varying parameters it covers a wide range of cosmological behaviours predicted by various other classes of models) and the lack of *a preferred* fundamental model make a good argument for using this model as a testbed, given that it is both *impossible* and *unnecessary* to study all chameleon-type models using simulations.

This chapter will be arranged as follows. In Sect. 6.2 we briefly describe the $f(R)$ gravity model and the chameleon screening mechanism. In Sect. 6.3 we recap the method currently employed in $f(R)$ simulations and explain why it is inefficient, before describing the new method. In Sect. 6.4 we perform some tests as validation of this new method. Finally, we discuss and summarise in Sect. 6.5.

In keeping with Chap. 5, throughout this chapter we follow the metric convention $(+, -, -, -)$, and set $c = 1$ except in the expressions where c appears explicitly. Greek indices $\mu, \nu, \ldots$ run over 0, 1, 2, 3. A subscript $_0$ denotes the present-day value of a quantity.

6.2   The Hu–Sawicki $f(R)$ Gravity Model

As we have already introduced the $f(R)$ model previously in this thesis (Sect. 5.2.1), the discussion in this section is kept brief. We remind the reader, however, that in the non-linear regime of structure formation in this model, assuming the quasistatic (Bose et al. 2015) and weak-field approximations, the modified Poisson equation is written as:

$$\nabla^2 \Phi = \frac{16\pi G}{3} \delta\rho_m - \frac{1}{6} \delta R\,(f_R),$$ (6.2.1)

which relates the gravitational potential Φ at a given position to the density ($\delta\rho_m \equiv \rho_m - \bar{\rho}_m$, where a bar denotes the cosmic mean of a quantity) and curvature ($\delta R \equiv R - \bar{R}$) at that position.

The equation of motion for the scalar field can be written as:

$$\nabla^2 f_R = \frac{1}{3c^2}\left[\delta R\,(f_R) - 8\pi G\delta\rho_m\right],$$ (6.2.2)

where $f_R = \mathrm{d}f(R)/\mathrm{d}R$.

Equations (6.2.1) and (6.2.2) need to be solved in cosmological simulations for $f(R)$ gravity to predict the modified gravitational force that is responsible for structure formation. It can be seen that Eq. (6.2.2) has a similar form to the Poisson equation, but $\delta R (f_R)$ is generally a non-linear function of f_R, and this makes it more difficult to numerically solve this equation.

Of course, to fully specify a $f(R)$ model one must fix the functional form $f(R)$. Without the guidance of a fundamental theory, it is not hard to imagine that there is no unique, or even preferred, way to do this. However, there are indeed practical considerations that mean that the functional form cannot be arbitrary either. This is because the choice of $f(R)$ must serve the purpose that it is originally designed for: namely, to explain the accelerated cosmic expansion. Moreover, as we shall see below, the design of $f(R)$ must ensure that any deviation from GR is suppressed to an insignificant level in places such as the Solar System, where numerous tests have confirmed compatibility with GR to high precision. Indeed, it is known (e.g., Refs. Brax et al. 2008, Wang et al. 2012a, Raveri et al. 2014, Ceron-Hurtado et al. 2016) that for any $f(R)$ model to pass Solar System gravity tests, the background evolution must be close to (practically indistinguishable from) that of ΛCDM.

The functional form of $f(R)$ we employ in this chapter is the same as that in Chap. 5, and is the one proposed by Hu and Sawicki (HS, Ref. Hu and Sawicki 2007), and has been shown to satisfy Solar System constraints. It is given as:

$$f(R) = -M^2 \frac{c_1 \left(-R/M^2\right)^n}{c_2 \left(-R/M^2\right)^n + 1},$$

(6.2.3)

where n, c_1, c_2 are dimensionless model parameters, and $M^2 = 8\pi G\rho_{m0}/3$ is another model parameter of mass dimension one that defines a characteristic mass scale for the theory. As in Chap. 5, we set $c_1/c_2 = 6\Omega_\Lambda/\Omega_m$ (in which Ω_m, Ω_Λ are, respectively, the present-day fractional density of non-relativistic matter and the cosmological constant), which guarantees that the model reproduces a ΛCDM expansion history at the background level.

The functional form of $f(R)$ is critical in determining if the fifth force can be sufficiently suppressed in dense environments. For the HS model, it was shown by Hu and Sawicki (2007) that $|f_{R0}| < 10^{-5}$ is required to screen the Milky Way, where f_{R0} is the background value of f_R today. Currently, the strongest constraint on the value of $|f_{R0}|$ in the HS model comes from the screening of dwarf galaxies, which requires $|f_{R0}| \lesssim 10^{-7}$ (95% C.L.) (Jain et al. 2013; Vikram et al. 2013). This is a promising way to constrain $f(R)$ gravity, provided astrophysical systematics are well controlled and the environmental impact on screening is modelled accurately (which itself will benefit from high resolution simulations).

In cosmology, there are many constraints on f_{R0} as well, and for recent reviews on this topic the readers are referred to Lombriser (2014), Burrage and Sakstein (2016). In Terukina et al. (2014), Wilcox (2015), X-ray and weak lensing estimates for the mass of the Coma cluster are combined to constrain $|f_{R0}| \lesssim 10^{-4.2}$ (95% C.L.). Two of the strongest constraints to date both come from cluster abundance.

In Cataneo (2015) the authors use X-ray cluster abundance while in Liu (2016) the counts of high-significance weak lensing convergence peaks are used as a proxy for cluster counts; both studies find that $|f_{R0}| \lesssim 10^{-5.2}$ after carefully accounting for systematics, even though the data and analyses are very different. In Cai et al. (2015), it was found that stacked lensing tangential shear of cosmic voids could potentially place constraints at a similar level. More recently, a study by Peirone et al. (2016), which uses Planck Sunyaev–Zel'dovich cluster counts, constrains $|f_{R0}| \lesssim 10^{-5.8}$, although the result is quite sensitive to the halo mass function used in the analysis. All the constraints are quoted at 95% confidence. There are many other cosmological and astrophysical constraints in the literature (e.g. from stellar evolution, Sakstein 2015), but it is beyond the scope of this chapter to mention all of them (some of these studies were carried out by using linear perturbation theory, which underestimates the effectiveness of screening and can therefore overestimate the strength of the constraints on the model – this is why simulations that fully capture the non-linearity of the theories are useful).

6.3 N-Body Equations and Algorithm

In this section, we describe the N-body equations in appropriate code units and their discretised versions that ECOSMOG solves in simulations.

6.3.1 The Newton–Gauss–Seidel Relaxation Method

Like its base code RAMSES (Teyssier 2002), ECOSMOG adopts supercomoving coordinates (Martel and Shapiro 1998) to express the field equations in terms of dimensionless quantities (see Eq. 5.4.33 for the full list). In terms of these variables, the Poisson and scalar field equations (Eqs. 6.2.1 and 6.2.2) in the HS model can be rewritten as:

$$\tilde{\nabla}^2 \tilde{\Phi} = 2\Omega_m (\tilde{\rho} - 1) - \frac{1}{6}\Omega_m a^4 \left[\left(-\frac{na^2 \xi}{\tilde{f}_R} \right)^{\frac{1}{n+1}} - 3\left(a^{-3} + 4\frac{\Omega_\Lambda}{\Omega_m} \right) \right], \quad (6.3.4)$$

$$\tilde{\nabla}^2 \tilde{f}_R = -\frac{1}{\tilde{c}^2}\Omega_m a(\tilde{\rho} - 1)$$

$$+ \frac{1}{3\tilde{c}^2}\Omega_m a^4 \left[\left(-\frac{na^2 \xi}{\tilde{f}_R} \right)^{\frac{1}{n+1}} - 3\left(a^{-3} + 4\frac{\Omega_\Lambda}{\Omega_m} \right) \right], \quad (6.3.5)$$

in which we have used the relation $m^2 = \Omega_m H_0^2$, and defined $\tilde{c} = c/(B H_0)$, which is the speed of light in code units. Note that these equations are the same as Eq. 5.4.35, but with the quasistatic approximation applied.

In principle, Eqs. (6.3.4) and (6.3.5) can be directly discretised on a mesh and can then be solved numerically. For chameleon-type models, however, there is a further subtlety: namely, the value of $-\tilde{f}_R$ is very small at early times and in high density regions. This property is desirable in order that the model can pass Solar System tests of gravity by virtue of the chameleon mechanism, but it also poses a challenge when solving Eq. (6.3.5) numerically. In the relaxation method that is employed to solve the discrete version of this equation, $-\tilde{f}_R$ in each mesh cell gets updated until the solution is close enough to the true value (more details below). This updating procedure is a numerical approximation, and it is possible that $-\tilde{f}_R$ can acquire negative numerical values in some cells as a result. Taking the case of the HS $n = 1$ model as an example: the quantity $(-\tilde{f}_R)^{\frac{1}{n+1}}$ is not physically defined if $-\tilde{f}_R < 0$, and the code then become unstable.

To overcome this numerical issue, in Oyaizu (2008) Oyaizu proposes to replace $-\tilde{f}_R$ with $\exp(u)$ in Eq. (6.3.5). As $\exp(u)$ can only be positive, this guarantees that the nonphysical situation described above will never appear. This change of variable has since then been used in all simulation codes of chameleon models to our knowledge (Oyaizu et al. 2008; Schmidt et al. 2009a; Zhao 2009, 2010; Zhao et al. 2010, 2011a; Li et al. 2012; Puchwein et al. 2013; Llinares et al. 2014; Wilcox et al. 2016).

In terms of this new variable, Eq. (6.3.5) can be discretised as:

$$
\frac{1}{h^2}\left[b_{i+\frac{1}{2},j,k}u_{i+1,j,k} - u_{i,j,k}\left(b_{i+\frac{1}{2},j,k} + b_{i-\frac{1}{2},j,k}\right) + b_{i-\frac{1}{2},j,k}u_{i-1,j,k}\right]
$$
$$
+\frac{1}{h^2}\left[b_{i,j+\frac{1}{2},k}u_{i,j+1,k} - u_{i,j,k}\left(b_{i,j+\frac{1}{2},k} + b_{i,j-\frac{1}{2},k}\right) + b_{i,j-\frac{1}{2},k}u_{i,j-1,k}\right]
$$
$$
+\frac{1}{h^2}\left[b_{i,j,k+\frac{1}{2}}u_{i,j,k+1} - u_{i,j,k}\left(b_{i,j,k+\frac{1}{2}} + b_{i,j,k-\frac{1}{2}}\right) + b_{i,j,k-\frac{1}{2}}u_{i,j,k-1}\right]
$$
$$
+\frac{1}{3\tilde{c}^2}\Omega_m a^4 \left(na^2\xi\right)^{\frac{1}{n+1}} \exp\left[-\frac{u_{i,j,k}}{n+1}\right] - \frac{1}{\tilde{c}^2}\Omega_m a(\rho_{i,j,k} - 1)
$$
$$
-\frac{1}{\tilde{c}^2}\Omega_m a^4 \left(a^{-3} + 4\frac{\Omega_\Lambda}{\Omega_m}\right) = 0, \tag{6.3.6}
$$

in which we have used the second order finite difference scheme to calculate $\tilde{\nabla}^2\left(-\tilde{f}_R\right)$. Defining the left-hand side of Eq. (6.3.6) as the operator $\mathscr{L}^h$, where a superscript ^{h} is used to denote that the equation is discretised on a mesh with cell size h, the equation can be written symbolically as:

$$
\mathscr{L}^h(u_{i,j,k}) = 0. \tag{6.3.7}
$$

This is a non-linear equation for $u_{i,j,k}$, and the most commonly used method to solve it is relaxation, which begins with some initial guesses of $u_{i,j,k}$ (for all mesh cells) and iteratively improves the old guess to get a new guess according to the Newton–Raphson method (same as the one used for solving non-linear algebraic equations):

$$u_{i,j,k}^{h,\text{new}} = u_{i,j,k}^{h,\text{old}} - \frac{\mathscr{L}^h\left(u_{i,j,k}^{h,\text{old}}\right)}{\frac{\partial \mathscr{L}^h\left(u_{i,j,k}^{h,\text{old}}\right)}{\partial u_{i,j,k}^h}}, \tag{6.3.8}$$

until $u_{i,j,k}$ (for all mesh cells) is close enough to the true solution or, equivalently, some all-mesh average of $\mathscr{L}^h\left(u_{i,j,k}\right)$ gets close enough to zero. A widely used definition of this all-mesh average (the so-called residual) is given by:

$$\text{Residual} \equiv \left[\sum_{i,j,k}\left[\mathscr{L}^h\left(u_{i,j,k}\right)\right]^2\right]^{1/2}, \tag{6.3.9}$$

where the summation is performed over all mesh cells on a given refinement level.

The implementation of this method is fairly straightforward in principle, but in practice there are a number of subtleties that need to be taken into account. For example, refined meshes usually have irregular shapes and their boundary conditions should be carefully set up by interpolating the values of u from coarser levels. The relaxation method is also notoriously slow to converge (convergence here meaning that the residual becomes smaller than some pre-fixed threshold) if it is only done on a fixed level, and in practice the so-called multigrid method is commonly used to remedy this (Brandt 1977). This consists of moving the equation to coarser meshes, solving it there, and then using the coarse-mesh solutions to correct the fine-mesh one. These subtleties have been discussed in detail in the literature; as they are not the main concern of this chapter, we refer interested readers to, e.g., Li et al. (2012), for a more elaborate description.

Although the multigrid method improves convergence in general, the rate of convergence is still very slow in $f(R)$ simulations, and the relaxation is some times unstable and diverges. One way to improve both the rate of convergence and the stability of the Newton–Gauss–Seidel relaxation method is to impose the condition:

$$\left|\mathscr{L}^h\left(u_{i,j,k}^{h,\text{new}}\right)\right| < \left|\mathscr{L}^h\left(u_{i,j,k}^{h,\text{old}}\right)\right|,$$

i.e., requiring that the residual after the new iteration gets monotonically smaller than in the previous one. When the condition is not met, we retain the value of the scalar field from the previous step ($u_{i,j,k}^{h,\text{old}}$). While satisfying this condition can be costly on each step, the overall efficiency of the code can be significantly increased by the improved numerical stability and reduced number of iterations required to reach convergence.

Finally, a similar discretisation can be done for the modified Poisson equation:

$$
\frac{1}{h^2}\left(\Phi_{i+1,j,k} + \Phi_{i-1,j,k} + \Phi_{i,j+1,k} + \Phi_{i,j-1,k} + \Phi_{i,j,k+1} + \Phi_{i,j,k-1} - 6\Phi_{i,j,k}\right)
$$
$$
= 2\Omega_m a(\rho_{i,j,k} - 1)
$$
$$
-\frac{1}{6}\Omega_m a^4\left[\left(na^2\xi\right)^{\frac{1}{n+1}}\exp\left(-\frac{u_{i,j,k}}{n+1}\right) - 3\left(a^{-3} + 4\frac{\Omega_\Lambda}{\Omega_m}\right)\right]. \tag{6.3.10}
$$

This equation is solved after Eq. (6.3.6), by which time $u_{i,j,k}$ is already known. As a result, this is a linear equation for $\Phi_{i,j,k}$ which is easier to solve than Eq. (6.3.6), and we shall not discuss it further here. Structurally, Eq. (6.3.10) is the same as the Poisson equation for standard gravity (with a modified source term); hence, one may simply use the standard RAMSES implementations for solving the Poisson equation.

6.3.2 The New Method

The discretisation used in the scalar field equation (Eq. 6.3.6) has a number of drawbacks:

- Depending on the value of ξ, the original scalar field equation can be very non-linear (when ξ is small, the term involving $\left(-\tilde{f}_R\right)^{\frac{1}{n+1}}$ is large and non-negligible, c.f. Eq. 5.2.12) or close to linear (when ξ is large, that term is small and negligible so that the equation becomes nearly linear in $\tilde{f}_R$).[4] In the former case, introducing the new variable $u = \log(-\tilde{f}_R)$ makes the equation even more non-linear; in the latter case, it non-linearises an almost linear equation. The high non-linearity makes the relaxation method very slow to converge, which is why simulations of $f(R)$ gravity are generally much more costly than ΛCDM simulations with the same specifications. Indeed, even with parallelised codes such as ECOSMOG, MG- GADGET, Isis and MG- ENZO (Zhao et al. in prep.), very large-sized and high resolution $f(R)$ simulations are currently still difficult to run, and this situation needs to be improved if we want to compare future survey data to theoretical predictions to perform accurate tests of modified gravity.
- As we have already seen above, the discrete Laplacian $\tilde{\nabla}^2 e^u$ is more complicated than the simple discretisation of $\tilde{\nabla}^2\tilde{\Phi}$, resulting in a more complex equation that needs to be solved.

---

[4]Note that, on first glance at Eq. (6.3.6), this may appear counter-intuitive. This dependence of the degree of linearity of Eq. (6.3.6) on the size of ξ can be explained by the fact that as ξ becomes smaller, the value of $\tilde{f}_R$ also becomes smaller (c.f. Eq. 5.2.12), making Eq. (6.3.6) on the whole more non-linear. The converse is true when ξ is large.

- The code ends up with a lot of exp and log operations. This is not optimal from a practical viewpoint, because the cost of these operations is generally much higher than that of simple arithmetic ones, such as summation, subtraction and multiplication.

The method described here alleviates the non-linearity problem by defining a new variable $u = \left(-\tilde{f}_R \right)^{1/2}$, so that the scalar field equation for the HS model with $n = 1$ (the most widely studied $f(R)$ model in the literature) becomes a simple cubic equation in u, which can be solved analytically instead of resorting to the approximation in Eq. (6.3.8):

$$u_{i,j,k}^3 + p u_{i,j,k} + q = 0, \tag{6.3.11}$$

where:

$$p \equiv \frac{h^2}{6\tilde{c}^2} \Omega_m a \tilde{\rho}_{i,j,k} + \frac{2h^2}{3\tilde{c}^2} \Omega_\Lambda a^4$$
$$- \frac{1}{6} \left(u_{i+1,j,k}^2 + u_{i-1,j,k}^2 + u_{i,j+1,k}^2 + u_{i,j-1,k}^2 + u_{i,j,k+1}^2 + u_{i,j,k-1}^2 \right), \tag{6.3.12}$$

$$q \equiv -\frac{h^2}{18\tilde{c}^2} \Omega_m a^4 \xi^{1/2}. \tag{6.3.13}$$

Note that here we focus on the case of $n = 1$; other cases will be discussed later.

While Eq. (6.3.11) can be solved analytically (and therefore accurately), it has three branches of solutions and, depending on the numerical values of p and q, all these branches can be real. Therefore, extra care has to be taken to make sure that the correct branch of solutions is chosen. For this, let us define:

$$\Delta_0 \equiv -3p,$$
$$\Delta_1 \equiv 27q. \tag{6.3.14}$$

As $q < 0$ is a constant in a given time step of the simulation, we see that $\Delta_1 < 0$.

The case $p > 0$ can occur in high density regions where $u > 0$ is small (and u^2 smaller still) because of the chameleon screening. In these cases, $\Delta_0 < 0$ and thus $\Delta_1^2 - 4\Delta_0^3 > 0$. The cubic equation then admits only one real solution, which must be the one we choose:

$$u_{i,j,k} = -\frac{1}{3} \left(C + \frac{\Delta_0}{C} \right) \tag{6.3.15}$$

with

$$C \equiv \left[\frac{1}{2} \left(\Delta_1 + \left(\Delta_1^2 - 4\Delta_0^3 \right)^{1/2} \right) \right]^{1/3}. \tag{6.3.16}$$

Note that Eq. (6.3.16) implies that $C = 0$ only when $\Delta_0 = p = 0$. This ensures that for the $p > 0$ case, $C \neq 0$ in Eq. (6.3.15), and the solution is never undefined.

In the case of $p = 0$, the solution is simply:

$$u_{i,j,k} = (-q)^{1/3}. \tag{6.3.17}$$

$p < 0$ can occur for density peaks in an overall low density region (where u and hence u^2 can be large). $\Delta_1^2 - 4\Delta_0^3$ can then take either positive or negative values. In the former case, the solution in Eq. (6.3.15) still holds, while in the latter case the equation has three real solutions:

$$u_{i,j,k} = -\frac{2}{3}\Delta_0^{1/2} \cos\left[\frac{1}{3}\Theta + \frac{2}{3}j\pi\right], \tag{6.3.18}$$

where $j = 0, 1, 2$ and $\cos\Theta \equiv \Delta_1 / \left(2\Delta_0^{3/2}\right)$. It is straightforward to decide which branch we should take: as $\Delta_1 < 0$, we have $\cos\Theta < 0$ and so $\Theta \in (\pi/2, \pi)$. Given that we require $u_{i,j,k}$ to be positive-definite:

- If $j = 0$, $u_{i,j,k} \sim -\cos\left(\frac{1}{3}\Theta\right) < 0$ and is unphysical;
- If $j = 1$, $u_{i,j,k} \sim -\cos\left(\frac{1}{3}\Theta + \frac{2}{3}\pi\right) > 0$ and is physical;
- If $j = 2$, $u_{i,j,k} \sim -\cos\left(\frac{1}{3}\Theta - \frac{2}{3}\pi\right) < 0$ and is unphysical.

This new method has a few interesting features:

- The discrete equation to be solved is significantly simpler. In particular, q is the same in all cells, so it only needs to be calculated once for a given time step and on a given mesh refinement level.
- There is a substantial reduction of costly computer operations as we get rid of operations. Some cos and $\cos^{-1}$ operations are introduced, but they will not be executed for all cells (depending on which branch of solutions we take); even for cells in which they need to be executed, they are only executed once. In the old method, exp is executed on both the cell and its neighbours.
- The cubic equation is solved analytically and a physical solution always exists. The variable redefinition in the old method, $\tilde{f}_R = \exp(u)$, was chosen so as to the avoid the unphysical solution $-\tilde{f}_R < 0$; the new method avoids this situation automatically by selecting the physical solution $u = \left(-\tilde{f}_R\right)^{1/2} > 0$ analytically. As a result, we can expect this new method to be both more stable (i.e., not suffering from catastrophic divergences due to numerics) and more efficient (i.e., the solution to Eq. (6.3.11) is exact for each Gauss–Seidel iteration, while Eq. (6.3.8) implicitly uses the approximate Newton–Raphson method and may need to be executed many times to arrive at what the new method achieves in one go).

Note that this new method does not really get rid of Gauss–Seidel relaxation, because the quantity p in Eq. (6.3.11) depends on the values of the scalar field in (the

6 direct) neighbouring cells, which are not accurate values but temporary guesses. It therefore still needs to do the Gauss–Seidel iterations (we use the standard red-black chessboard scheme here). What it does get rid of is the 'Newton–Raphson' part (Eq. (6.3.8)) of the Newton–Gauss–Seidel (or non-linear GS, or NGS) relaxation which updates the old guesses using a linear approximation of the full non-linear equation. The speedup is also largely assisted by the simplicity of Eq. (6.3.12) compared to Eq. (6.3.6), which comes about due to the new variable redefinition. Therefore, while Gauss–Seidel iterations are still required, the savings using the new method can still be significant.

6.4 Tests and Simulations of the New Method

In this section we present the results of several test runs of the new ECOSMOG code. In what follows, we will only consider the F6 model of $f(R)$ gravity, in which the present-day value of the scalar field is given by $\left|\bar{f}_{R0}\right| = 10^{-6}$. In this model, the chameleon screening is particularly efficient, meaning that deviations from GR are very small. To capture the effects of screening, accurately solving the non-linear scalar field equations is therefore necessary.

We have simulated the F6 model at three resolution levels: 'Low res', 'Medium res' and 'High res' (the box size and number of particles used in each of these runs are summarised in Table 6.1). In each case, we have also run a ΛCDM simulation starting from the same initial conditions. The mesh refinement criteria used for the 'High res' simulation allows us to resolve small scales comparable to those in the Millennium simulation (Springel et al. 2005). While the 'Low' and 'High res' runs use Planck 2015 (Planck Collaboration et al. 2016) cosmological parameters (with $\Omega_m = 0.308, \Omega_\Lambda = 0.692, h = 0.6781, \sigma_8 = 0.8149$), the cosmological parameters for the 'Medium res' run are obtained from WMAP-7 (Komatsu et al. 2011) data (with $\Omega_m = 0.271, \Omega_\Lambda = 0.729, h = 0.704, \sigma_8 = 0.8092$).

Table 6.1 Details of the simulations performed in this work. The columns B and N_p, respectively, refer to the comoving box size and number of particles in each of these runs. The starting redshift in all simulations was $z_{ini} = 49$. The second last column summarises the factor by which the new method is faster than the old one in each case. Note that the $>20\times$ speedup for the 'High res' simulation is an estimate - we have not run an F6 simulation at this resolution using the old method. The last column shows the percentage overhead of the F6 simulations using the new method compared to ΛCDM. The level of speedup that can be achieved in the F6 simulations depends on the convergence criteria used: in all cases, convergence is considered as achieved when the residual is $<10^{-8}$ on the domain level, and $<10^{-7}$ on the fine levels

Name	Model	B [Mpc/h]	N_p	Speed up	Overhead (new method) (%)
Low res	ΛCDM, F6	512	512^3	$5\times$	110
Medium res	ΛCDM, F6	250	512^3	$15\times$	180
High res	ΛCDM, F6	128	512^3	$>20\times$	190

Fig. 6.1 *Top panel*:
Comparison of the
non-linear matter power
spectra at $z = 0, 0.5$ for the
F6 model using the old
method (blue, Sect. 6.3.1)
and the new method (red,
Sect. 6.3.2) for solving the
scalar field equations of
motion. The results shown
are for the 'Medium res'
simulation. *Bottom panel*:
Ratio of the power spectra
corresponding to the upper
panel. The shaded grey band
represents a 1% error region.
Figure reproduced from
Bose et al. (2017)

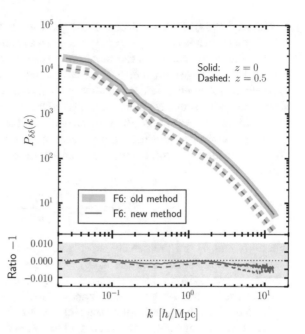

In Fig. 6.1, we compare the non-linear matter power spectrum, $P_{\delta\delta}(k)$, from the 'Medium res' simulations using the old and new methods. $P_{\delta\delta}(k)$ was computed using the publicly-available POWMES code (Colombi et al. 2009). The solid and dashed curves are $P_{\delta\delta}(k)$ computed at $z = 0$ and $z = 0.5$, respectively, for F6. The results of the two methods are indistinguishable at both redshifts, and this is quantified more clearly in the bottom panel of Fig. 6.1, which shows the relative difference between the old and new methods. The shaded grey band in this panel represents a 1% error around zero; clearly, the new and old methods agree to well below 1% at all scales resolved in the simulation. The same is true even at higher redshift ($z = 1, 2$, not shown). We have checked that the agreement also holds in the case of the velocity divergence power spectrum, $P_{\theta\theta}(k)$, which, being just the first integral of the gravitational acceleration, would be more sensitive to differences in the gravitational forces between the two methods. Agreement for $P_{\theta\theta}(k)$, which is calculated in a volume-weighted way, shows that the two methods agree well even in regions of the cosmic web that are not mass-dominated. This is not unexpected: after all, the new method solves the same equation of motion, without needing to use the approximate and inefficient Newton–Raphson scheme. As a consequence, the simulation is now *significantly* faster than before: the new method boosts the speed of the F6 calculation by a factor of 15 relative to the old implementation in ECOSMOG (see the last column of Table 6.1).

Two-point statistics such as the power spectrum offer a complete description of clustering properties only for Gaussian fields. Gravitational instability theory predicts that the non-linear evolution induced by gravity drives away the PDF of these fields from Gaussianity at late times and small scales (see e.g., Juszkiewicz et al. 1993,

Bernardeau 1994). This is reflected in the growing skewness and kurtosis of cosmic
density and velocity fields. $f(R)$ theories show systematic deviations from ΛCDM
for these statistics, and these can therefore be used as a test of the theory (Hellwing
et al. 2013). We have computed PDFs and their higher-order moments for the density
and velocity divergence fields to test how well the old and new methods agree beyond
simple two-point statistics. We find that the differences are very small and comparable
to the differences seen in the $P(k)$. As an additional test, we have also computed the
Fourier mode decoherence functions (Strauss et al. 1992; Chodorowski and Ciecielag
2002), defined as Pearson correlation coefficients for the Fourier modes of the two
fields:

$$ C(k) \equiv \frac{\langle f_1 f_2^* \rangle}{\langle f_1^2 \rangle^{1/2} \langle f_2^2 \rangle^{1/2}}, $$

where f_1 and f_2 are the density or velocity divergence fields for the $f(R)$ runs
computed using the two methods. $C(k) = 1$ when both fields being compared have
Fourier modes at given k that correspond exactly. The density and velocity divergence
fields for the F6 runs using the two methods take $C(k) = 1$ for almost the entire range
of k, up until the Nyquist limit of the simulations. These tests reassure us that the
density and velocity fields produced by the old and the new method are, for all
practical purposes, indistinguishable.

Results from the 'High res' simulations are shown in Fig. 6.2, where we plot the
relative difference in $P_{\delta\delta}(k)$ of F6 with ΛCDM – only results using the new method

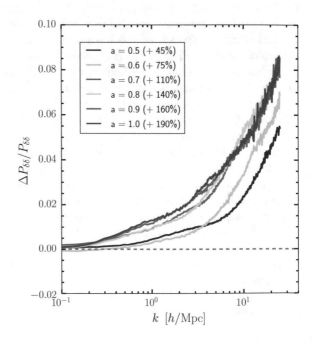

Fig. 6.2 Enhancement of
the F6 matter power
spectrum relative to ΛCDM
for the 'High res' simulation
($B = 128\,\mathrm{Mpc/h}$,
$N_p = 512^3$). The different
coloured curves show the
relative difference at
different scale factors, as
indicated in the legend.
Alongside the legend labels,
we also note the percentage
overhead associated with the
F6 run compared to the
ΛCDM run at the same scale
factor. Figure reproduced
from Bose et al. (2017)

are shown. Curves of different colours represent the relative difference at different scale factors, as labelled in the legend. The legend labels also list the percentage overhead involved in the F6 simulation compared to the ΛCDM run at the same scale factor. With the new method, the F6 simulation is now only $\sim$45% slower than the ΛCDM run at $a = 0.5$ ($z = 1$), and only $\sim$190% slower at the final time. Compared to F6 simulations with comparable resolution using the old method (e.g., Ref. Shi et al. 2015), the new implementation is estimated to be *more than* 20 $\times$ *faster*.

The degree to which the new method improves the efficiency of ECOSMOG over the old one depends on resolution. Indeed, in going from the 'Low res' to the 'High res' simulations, the gain in performance increases from a factor of 5 to a factor of over 20 (the overhead increases considerably with resolution in the old method). The improved efficiency of the numerical algorithm will enable us to run simulations of chameleon models that would previously have been computationally very expensive to perform. Future applications of the method could include running hydrodynamical simulations (where high resolution is required to follow accurately the hydrodynamics and to resolve spatial scales important for star formation and feedback), and running large numbers of low resolution volumes to estimate the covariance matrix in non-standard gravity.

6.5 Summary and Discussions

Modified gravity models are an umbrella group of theories seeking to explain the apparent accelerated cosmic expansion by assuming modifications to the Einsteinian gravitational law on cosmological scales. Usually, such modifications must be small in high density environments in which gravity is known to be accurately described by GR, and this can be achieved by screening mechanisms, resulting in highly non-linear field equations. Studying the cosmological implications of these theories and observational constraints on them is an active research topic in cosmology, but the non-linear nature of these theories means that one has to resort to numerical simulations, which can be prohibitively slow. This has, up until now, limited the scope of accurately testing gravity using precision observational data.

In this chapter, we proposed and demonstrated the power of a new and more efficient method to solve the non-linear field equation in one of the most popular modified gravity models – the Hu–Sawicki variant of $f(R)$ gravity. The current method used to simulate this model is slow mainly because of a variable redefinition aimed at making the relaxation algorithm numerically stable, but has the negative side effect of making the discrete equation even more non-linear and, therefore, harder to converge. As a result, modified gravity simulations which match the size *and* resolution of the state-of-the-art ΛCDM N-body or hydrodynamical simulations have thus far been beyond reach (but see Hammami et al. 2015, Arnold et al. 2016).

The new method avoids the specific variable redefinition used in the old method, and therefore does not further increase the non-linearity of the discrete equation to be solved. More importantly, it enables the discrete equation to be written in a form that is *analytically* solvable at each Gauss–Seidel iteration. This is what ultimately makes the method efficient: compare solving a highly non-linear algebraic equation analytically and solving the same equation using the Newton–Raphson iteration method (Eq. 6.3.8), and it is clear that the latter is generally much more inefficient.

We have performed test simulations using the new method, and confirmed that it is indeed very efficient. The working model for the tests is the F6 variant of Hu–Sawicki $f(R)$ gravity. The chameleon screening is very efficient in F6, and it is therefore important that the non-linear scalar field equations are solved accurately. In Fig. 6.1, we have confirmed that the new and old methods agree at the sub-percent level when comparing the non-linear matter power spectrum, $P_{\delta\delta}(k)$. The good agreement continues to hold at higher redshift, as well as for the velocity divergence power spectra, $P_{\theta\theta}(k)$. Next, in Fig. 6.2, we presented results from our 'High res' simulations, which are comparable in resolution to the Millennium simulation. The total overhead in the F6 simulation is $\sim$190% compared to the equivalent ΛCDM run; this represents a boost in performance of $>20\times$ compared to an F6 simulation of similar resolution using the old method.

The improved performance of the new simulation algorithm compared to the old one serves to highlight the importance of the way in which one discretises partial differential equations for the efficiency of numerically solving them. This is particularly true for highly non-linear equations, such as those encountered in many modified gravity theories. Our work highlights the following:

(1) There is not a single way of discretisation, and this usually depends on the specific equations to be solved. In general, the discretisation should be chosen to preserve the original degree of non-linearity of the equation as much as possible, and avoid further non-linearising the equation.

(2) Where possible, exact solutions to the non-linear discrete equation should be used instead of the approximate solution in Eq. (6.3.8). The latter, despite being commonly used in relaxation solutions to non-linear differential equation, is a second option only for cases where $\mathscr{L}^h\left(u_{i,j,k}\right) = 0$ has no analytical solution in general.

The same observations and conclusions apply to other classes of partial differential equations, such as those involving higher order powers of the derivatives of the scalar field (e.g., $\left(\nabla^2\varphi\right)^2$, $\nabla^i\nabla^j\varphi\nabla_i\nabla_j\varphi$, $\left(\nabla^2\varphi\right)^3$, $\nabla^i\nabla_j\varphi\nabla^j\nabla_k\varphi\nabla^k\nabla_i\varphi$), which are commonly encountered in Vainshtein-type theories. In fact, in the most popular examples of such models – the dgp, cubic Galileon and quartic Galileon models – we also found that the discretisation could be done in a way such that $\mathscr{L}^h\left(u_{i,j,k}\right) = 0$ is a quadratic or cubic equation that can be solved analytically. This fact has been used in Li et al. (2013a); Li et al. (2013a), Barreira et al. (2013); Barreira et al. (2015) to make simulations of these models possible, more efficient and free from numerical instabilities.

Unfortunately, the new method does *not* apply to *all* non-linear partial differential equations, because it relies on $u_{i,j,k}$ being analytically solvable in the discrete equation. In the HS $f(R)$ model with $n = 1$, $u_{i,j,k}$ satisfies a cubic equation, which does have analytical solutions. This neat property does not hold for other models. However, this method will still be very useful for the following reasons:

- At the moment, no specific functional forms of $f(R)$ – or more generally, no specific chameleon models – are known to be fundamental. Different models often share similar qualitative behaviours though the predictions can be quantitatively different. For what it is worth, the HS model serves as a great test case to gain insights into the question 'How much deviation from GR (in the manner prescribed by the large class of chameleon models) is allowed by cosmological data?'. Indeed, all current observational constraints on modified gravity are to be considered as attempts to answer this question. In this context, the exact functional form of $f(R)$ is not critical, because whatever form we adopt, it is unlikely to be the true theory. Actually, the HS model is capable of reproducing the behaviours of many classes of models, and is therefore a representative example.
- There are other models that this method can be applied to. One example is the HS $f(R)$ model with $n = 2$. In this case Eq. (6.3.11) becomes a quartic equation, which also has analytical solutions. A further example is the logarithmic $f(R)$ model studied in the literature (e.g., Ref. Brax et al. 2008):

$$f(R) \sim -2\Lambda - \eta \log (R/R_*),$$

where Λ is the cosmological constant, and η and R_* are some model parameters. In this case, $f_R \sim 1/R$, and we could define $u = -\tilde{f}_R$ so that Eq. (6.3.11) becomes a quadratic equation. Moreover, looking beyond $f(R)$ gravity, there are also other chameleon models with different coupling strengths from the value of $1/3$ for $f(R)$ models, and can be simulated using this method (Brax et al. 2013). The method can also be applied to the symmetron model (Hinterbichler and Khoury 2010), in which the equation:

$$\mathscr{L}^h \left(\varphi_{i,j,k} \right) = 0,$$

is a cubic equation (Davis et al. 2012) for the symmetron field φ, and certain variants of the dilaton model (Brax et al. 2010, 2012a), though our initial tests showed that the improvement of the efficiency is far smaller than in the $f(R)$ case (Appendix A.2).

Efforts towards generalising the new method to the models mentioned above, and to running large high resolution simulations including baryonic physics, are currently ongoing and will be the subject of future works.

References

Ali A, Gannouji R, Sami M (2010) Phys Rev D 82:103015. https://doi.org/10.1103/PhysRevD.82.103015, http://adsabs.harvard.edu/abs/2010PhRvD..82j3015A

Arnold C, Springel V, Puchwein E (2016) MNRAS 462:1530. https://doi.org/10.1093/mnras/stw1708, http://adsabs.harvard.edu/abs/2016MNRAS.462.1530A

Barreira A, Li B, Baugh CM, Pascoli S (2012) Phys Rev D 86:124016. https://doi.org/10.1103/PhysRevD.86.124016, http://adsabs.harvard.edu/abs/2012PhRvD..86l4016B

Barreira A, Li B, Hellwing WA, Baugh CM, Pascoli S (2013) JCAP 10:027. https://doi.org/10.1088/1475-7516/2013/10/027, http://adsabs.harvard.edu/abs/2013JCAP...10..027B

Barreira A, Bose S, Li B (2015) JCAP 12:059. https://doi.org/10.1088/1475-7516/2015/12/059, http://adsabs.harvard.edu/abs/2015JCAP...12..059B

Barreira A, Bose S, Li B, Llinares C (2017) JCAP 2:031. https://doi.org/10.1088/1475-7516/2017/02/031, http://adsabs.harvard.edu/abs/2017JCAP...02..031B

Bernardeau F (1994) ApJ 433:1. https://doi.org/10.1086/174620, http://adsabs.harvard.edu/abs/1994ApJ...433....1B

Beutler F et al (2011) MNRAS 416:3017. https://doi.org/10.1111/j.1365-2966.2011.19250.x, http://adsabs.harvard.edu/abs/2011MNRAS.416.3017B

Bose S, Hellwing WA, Li B (2015) JCAP 2:034. https://doi.org/10.1088/1475-7516/2015/02/034, http://adsabs.harvard.edu/abs/2015JCAP...02..034B

Bose S, Li B, Barreira A, He J-h, Hellwing WA, Koyama K, Llinares C, Zhao G-B (2017) JCAP 2:050. https://doi.org/10.1088/1475-7516/2017/02/050, http://adsabs.harvard.edu/abs/2017JCAP...02..050B

Brandt A (1977) Math Comput 31:333

Brax P, Burrage C, Davis A-C (2011) JCAP 9:020. https://doi.org/10.1088/1475-7516/2011/09/020, http://adsabs.harvard.edu/abs/2011JCAP...09..020B

Brax P, van de Bruck C, Davis A-C, Shaw DJ (2008) Phys Rev D 78:104021. https://doi.org/10.1103/PhysRevD.78.104021, http://adsabs.harvard.edu/abs/2008PhRvD..78j4021B

Brax P, van de Bruck C, Davis A-C, Shaw D (2010) Phys Rev D 82:063519. https://doi.org/10.1103/PhysRevD.82.063519

Brax P, Davis A-C, Li B, Winther HA, Zhao G-B (2012a) JCAP 10:002. https://doi.org/10.1088/1475-7516/2012/10/002, http://adsabs.harvard.edu/abs/2012JCAP...10..002B

Brax P, Davis A-C, Li B, Winther HA (2012b) Phys Rev D 86:044015. https://doi.org/10.1103/PhysRevD.86.044015, http://adsabs.harvard.edu/abs/2012PhRvD..86d4015B

Brax P, Davis A-C, Li B (2012c) Phys Lett B 715:38. https://doi.org/10.1016/j.physletb.2012.08.002, http://adsabs.harvard.edu/abs/2012PhLB..715...38B

Brax P, Davis A-C, Li B, Winther HA, Zhao G-B (2013) JCAP 4:029. https://doi.org/10.1088/1475-7516/2013/04/029, http://adsabs.harvard.edu/abs/2013JCAP...04..029B

Brax P, Valageas P (2014a) Phys Rev D 90:023507. https://doi.org/10.1103/PhysRevD.90.023507, http://adsabs.harvard.edu/abs/2014PhRvD..90b3507B

Brax P, Valageas P (2014b) Phys Rev D 90:023508 https://doi.org/10.1103/PhysRevD.90.023508, http://adsabs.harvard.edu/abs/2014PhRvD..90b3508B

Burrage C, Sakstein J (2016) JCAP 11:045. https://doi.org/10.1088/1475-7516/2016/11/045, http://adsabs.harvard.edu/abs/2016JCAP...11..045B

Cai Y-C, Padilla N, Li B (2015) MNRAS 451:1036. https://doi.org/10.1093/mnras/stv777, http://adsabs.harvard.edu/abs/2015MNRAS.451.1036C

Cataneo M et al (2015) Phys Rev D 92:044009. https://doi.org/10.1103/PhysRevD.92.044009, http://adsabs.harvard.edu/abs/2015PhRvD..92d4009C

Cataneo M, Rapetti D, Lombriser L, Li B (2016) JCAP 12:024. https://doi.org/10.1088/1475-7516/2016/12/024, http://adsabs.harvard.edu/abs/2016JCAP...12..024C

Ceron-Hurtado JJ, He J-h, Li B (2016) Phys Rev D 94:064052. https://doi.org/10.1103/PhysRevD.94.064052, http://adsabs.harvard.edu/abs/2016PhRvD..94f4052C

Chkareuli G, Pirtskhalava D (2012) Phys Lett B 713:99. https://doi.org/10.1016/j.physletb.2012.05.030, http://adsabs.harvard.edu/abs/2012PhLB..713...99C

Chodorowski MJ, Ciecielag P (2002) MNRAS 331:133. https://doi.org/10.1046/j.1365-8711.2002.05161.x, http://adsabs.harvard.edu/abs/2002MNRAS.331..133C

Chow N, Khoury J (2009) Phys Rev D 80:024037. https://doi.org/10.1103/PhysRevD.80.024037, http://adsabs.harvard.edu/abs/2009PhRvD..80b4037C

Clifton T, Ferreira PG, Padilla A, Skordis C (2012) Phys Rep 513:1. https://doi.org/10.1016/j.physrep.2012.01.001, http://adsabs.harvard.edu/abs/2012PhR...513....1C

Colombi S, Jaffe A, Novikov D, Pichon C (2009) MNRAS 393:511. https://doi.org/10.1111/j.1365-2966.2008.14176.x, http://adsabs.harvard.edu/abs/2009MNRAS.393..511C

Copeland EJ, Sami M, Tsujikawa S (2006) Int J Mod Phys D 15:1753. https://doi.org/10.1142/S021827180600942X, http://adsabs.harvard.edu/abs/2006IJMPD..15.1753C

Davis A-C, Li B, Mota DF, Winther HA (2012) ApJ 748:61. https://doi.org/10.1088/0004-637X/748/1/61, http://adsabs.harvard.edu/abs/2012ApJ...748...61D

de Rham C, Gabadadze G, Tolley AJ (2011) Phys Rev Lett 106:231101. https://doi.org/10.1103/PhysRevLett.106.231101, http://adsabs.harvard.edu/abs/2011PhRvL.106w1101D

Deffayet C, Esposito-Farèse G, Vikman A (2009) Phys Rev D 79:084003. https://doi.org/10.1103/PhysRevD.79.084003, http://adsabs.harvard.edu/abs/2009PhRvD..79h4003D

DESI Collaboration et al (2016a). arXiv:1611.00036

DESI Collaboration et al (2016b). arXiv:1611.00037

Dvali G, Gabadadze G, Porrati M (2000) Phys Lett B 485:208. https://doi.org/10.1016/S0370-2693(00)00669-9, http://adsabs.harvard.edu/abs/2000PhLB..485..208D

Falck B, Koyama K, Zhao G-B (2015) JCAP 7:049. https://doi.org/10.1088/1475-7516/2015/07/049, http://adsabs.harvard.edu/abs/2015JCAP...07..049F

Guy J et al (2010) A& A. https://doi.org/10.1051/0004-6361/201014468, http://adsabs.harvard.edu/abs/2010A

Hammami A, Llinares C, Mota DF, Winther HA (2015) MNRAS 449:3635. https://doi.org/10.1093/mnras/stv529, http://adsabs.harvard.edu/abs/2015MNRAS.449.3635H

Hellwing WA, Li B, Frenk CS, Cole S (2013) MNRAS 435:2806. https://doi.org/10.1093/mnras/stt1430, http://adsabs.harvard.edu/abs/2013MNRAS.435.2806H

Hinshaw G et al (2013) APJS 280:19. https://doi.org/10.1088/0067-0049/208/2/19, http://adsabs.harvard.edu/abs/2013ApJS..208...19H

Hinterbichler K, Khoury J (2010) Phys Rev Lett 104:231301. https://doi.org/10.1103/PhysRevLett.104.231301, http://adsabs.harvard.edu/abs/2010PhRvL.104w1301H

Hu W, Sawicki I (2007) Phys Rev D 76:064004. https://doi.org/10.1103/PhysRevD.76.064004, http://adsabs.harvard.edu/abs/2007PhRvD..76f4004H

Ivezic Z et al (2008). arXiv:0805.2366

Jain B, Vikram V, Sakstein J (2013) ApJ 779:39. https://doi.org/10.1088/0004-637X/779/1/39, http://adsabs.harvard.edu/abs/2013ApJ...779...39J

Jennings E, Baugh CM, Li B, Zhao G-B, Koyama K (2012) MNRAS 425:2128. https://doi.org/10.1111/j.1365-2966.2012.21567.x, http://adsabs.harvard.edu/abs/2012MNRAS.425.2128J

Joyce A, Lombriser L, Schmidt F (2016) Ann Rev Nucl Part Sci 66:95. https://doi.org/10.1146/annurev-nucl-102115-044553, http://adsabs.harvard.edu/abs/2016ARNPS..66...95J

Juszkiewicz R, Bouchet FR, Colombi S (1993) ApJL 412:L9. https://doi.org/10.1086/186927, http://adsabs.harvard.edu/abs/1993ApJ...412L...9J

Khoury J, Weltman A (2004) Phys Rev D 69:044026. https://doi.org/10.1103/PhysRevD.69.044026, http://adsabs.harvard.edu/abs/2004PhRvD..69d4026K

Komatsu E et al (2011) APJS 192:18. https://doi.org/10.1088/0067-0049/192/2/18, http://adsabs.harvard.edu/abs/2011ApJS..192...18K

Koyama K (2016) Rep Prog Phys 79:046902. https://doi.org/10.1088/0034-4885/79/4/046902, http://adsabs.harvard.edu/abs/2016RPPh...79d6902K

Laureijs R et al (2011). arXiv:1110.3193

Lee J, Zhao G-B, Li B, Koyama K (2013) ApJ 763:28. https://doi.org/10.1088/0004-637X/763/1/28, http://adsabs.harvard.edu/abs/2013ApJ...763...28L

Levi M et al (2013). arXiv:1308.0847

Li B, Barrow JD (2007) Phys Rev D 75:084010. https://doi.org/10.1103/PhysRevD.75.084010, http://adsabs.harvard.edu/abs/2007PhRvD..75h4010L

Li B, Efstathiou G (2012) MNRAS 421:1431. https://doi.org/10.1111/j.1365-2966.2011.20404.x, http://adsabs.harvard.edu/abs/2012MNRAS.421.1431L

Li Y, Hu W (2011) Phys Rev D 84:084033. https://doi.org/10.1103/PhysRevD.84.084033, http://adsabs.harvard.edu/abs/2011PhRvD..84h4033L

Li B, Zhao H (2009) Phys Rev D 80:044027. https://doi.org/10.1103/PhysRevD.80.044027, http://adsabs.harvard.edu/abs/2009PhRvD..80d4027L

Li B, Zhao H (2010) Phys Rev D 81:104047. https://doi.org/10.1103/PhysRevD.81.104047, http://adsabs.harvard.edu/abs/2010PhRvD..81j4047L

Li B, Zhao G-B, Teyssier R, Koyama K (2012a) JCAP 1:051. https://doi.org/10.1088/1475-7516/2012/01/051, http://adsabs.harvard.edu/abs/2012JCAP...01..051L

Li B, Zhao G-B, Koyama K (2012b) MNRAS 421:3481. https://doi.org/10.1111/j.1365-2966.2012.20573.x, http://adsabs.harvard.edu/abs/2012MNRAS.421.3481L

Li B, Zhao G-B, Koyama K (2013a) JCAP 5:025. https://doi.org/10.1088/1475-7516/2013/05/023, http://adsabs.harvard.edu/abs/2013JCAP...05..023L

Li B, Barreira A, Baugh CM, Hellwing WA, Koyama K, Pascoli S, Zhao G-B (2013) [JCAP] 11:012. https://doi.org/10.1088/1475-7516/2013/11/012, http://adsabs.harvard.edu/abs/2013JCAP...11..012L

Li B, Hellwing WA, Koyama K, Zhao G-B, Jennings E, Baugh CM (2013c) MNRAS 428:743. https://doi.org/10.1093/mnras/sts072, http://adsabs.harvard.edu/abs/2013MNRAS.428..743L

Liu X et al (2016) Phys Rev Lett 117:051101. https://doi.org/10.1103/PhysRevLett.117.051101, http://adsabs.harvard.edu/abs/2016PhRvL.117e1101L

Llinares C, Knebe A, Zhao H (2008) MNRAS 391:1778. https://doi.org/10.1111/j.1365-2966.2008.13961.x, http://adsabs.harvard.edu/abs/2008MNRAS.391.1778L

Llinares C, Mota DF, Winther HA (2014) A&A. https://doi.org/10.1051/0004-6361/201322412, http://adsabs.harvard.edu/abs/2014A

Lombriser L (2014) Ann der Phys 526:259. https://doi.org/10.1002/andp.201400058, http://adsabs.harvard.edu/abs/2014AnP...526..259L

Lombriser L, Koyama K, Zhao G-B, Li B (2012) Phys Rev D 85:124054. https://doi.org/10.1103/PhysRevD.85.124054, http://adsabs.harvard.edu/abs/2012PhRvD..85l4054L

Martel H, Shapiro PR (1998) MNRAS 297:467. https://doi.org/10.1046/j.1365-8711.1998.01497.x, http://adsabs.harvard.edu/abs/1998MNRAS.297..467M

Mead AJ, Peacock JA, Lombriser L, Li B (2015) MNRAS 452:4203.https://doi.org/10.1093/mnras/stv1484, http://adsabs.harvard.edu/abs/2015MNRAS.452.4203M

Merloni A et al (2012). arXiv:1209.3114

Mota DF, Shaw DJ (2007) Phys Rev D 75:063501. https://doi.org/10.1103/PhysRevD.75.063501, http://adsabs.harvard.edu/abs/2007PhRvD..75f3501M

Neveu J, Ruhlmann-Kleider V, Astier P, Besançon M, Guy J, Möller A, Babichev E (2017) A&A. https://doi.org/10.1051/0004-6361/201628878, http://adsabs.harvard.edu/abs/2017A

Nicolis A, Rattazzi R, Trincherini E (2009) Phys Rev D 79:064036. https://doi.org/10.1103/PhysRevD.79.064036, http://adsabs.harvard.edu/abs/2009PhRvD..79f4036N

Oyaizu H (2008) Phys Rev D 78:123523. https://doi.org/10.1103/PhysRevD.78.123523, http://adsabs.harvard.edu/abs/2008PhRvD..78l3523O

Oyaizu H, Lima M, Hu W (2008) Phys Rev D 78:123524. https://doi.org/10.1103/PhysRevD.78.123524, http://adsabs.harvard.edu/abs/2008PhRvD..78l3524O

Peirone S, Raveri M, Viel M, Borgani S, Ansoldi S (2016). http://adsabs.harvard.edu/abs/2016arXiv160707863P, arXiv:1607.07863

Percival WJ et al (2010) MNRAS 401:2148. https://doi.org/10.1111/j.1365-2966.2009.15812.x, http://adsabs.harvard.edu/abs/2010MNRAS.401.2148P

Planck Collaboration et al (2016) A&A. https://doi.org/10.1051/0004-6361/201525830, http://adsabs.harvard.edu/abs/2016A

Puchwein E, Baldi M, Springel V (2013) MNRAS 436:348. https://doi.org/10.1093/mnras/stt1575, http://adsabs.harvard.edu/abs/2013MNRAS.436..348P

Raveri M, Hu B, Frusciante N, Silvestri A (2014) Phys Rev D 90:043513. https://doi.org/10.1103/PhysRevD.90.043513, http://adsabs.harvard.edu/abs/2014PhRvD..90d3513R

Reid BA et al (2012) MNRAS 426:2719. https://doi.org/10.1111/j.1365-2966.2012.21779.x, http://adsabs.harvard.edu/abs/2012MNRAS.426.2719R

Riess AG et al (2009) ApJ 699:539. https://doi.org/10.1088/0004-637X/699/1/539, http://adsabs.harvard.edu/abs/2009ApJ...699..539R

Sakstein J (2015) Phys Rev D 92:124045. https://doi.org/10.1103/PhysRevD.92.124045, http://adsabs.harvard.edu/abs/2015PhRvD..92l4045S

Sbisà F, Niz G, Koyama K, Tasinato G (2012) Phys Rev D 86:024033. https://doi.org/10.1103/PhysRevD.86.024033, http://adsabs.harvard.edu/abs/2012PhRvD..86b4033S

Schmidt F, Lima M, Oyaizu H, Hu W (2009) Phys Rev D 79:083518. https://doi.org/10.1103/PhysRevD.79.083518, http://adsabs.harvard.edu/abs/2009PhRvD..79h3518S

Shi D, Li B, Han J, Gao L, Hellwing WA (2015) MNRAS 452:3179. https://doi.org/10.1093/mnras/stv1549, http://adsabs.harvard.edu/abs/2015MNRAS.452.3179S

Silva FP, Koyama K (2009) Phys Rev D 80:121301. https://doi.org/10.1103/PhysRevD.80.121301, http://adsabs.harvard.edu/abs/2009PhRvD..80l1301S

Sotiriou TP, Faraoni V (2010) Rev Mod Phys 82:451. https://doi.org/10.1103/RevModPhys.82.451, http://adsabs.harvard.edu/abs/2010RvMP...82..451S

Springel V et al (2005) Nature 435:629. https://doi.org/10.1038/nature03597, http://adsabs.harvard.edu/abs/2005Natur.435..629S

Strauss MA, Yahil A, Davis M, Huchra JP, Fisher K (1992) ApJ 397:395. https://doi.org/10.1086/171796, http://adsabs.harvard.edu/abs/1992ApJ...397..395S

Ternkina A, Lombriser L, Yamamoto K, Bacon D, Koyama K, Nichol RC (2014) JCAP 4:013. https://doi.org/10.1088/1475-7516/2014/04/013, http://adsabs.harvard.edu/abs/2014JCAP...04..013T

Teyssier R (2002) A & A. https://doi.org/10.1051/0004-6361:20011817, http://adsabs.harvard.edu/abs/2002A

Vainshtein A (1972) Phys Lett B 39:393. https://doi.org/10.1016/0370-2693(72)90147-5

Vikram V, Cabré A, Jain B, VanderPlas JT (2013) JCAP 8:020. https://doi.org/10.1088/1475-7516/2013/08/020, http://adsabs.harvard.edu/abs/2013JCAP...08..020V

Wang J, Hui L, Khoury J (2012) Phys Rev Lett 109:241301. https://doi.org/10.1103/PhysRevLett.109.241301, http://adsabs.harvard.edu/abs/2012PhRvL.109x1301W

Wilcox H et al (2015) MNRAS 452:1171. https://doi.org/10.1093/mnras/stv1366, http://adsabs.harvard.edu/abs/2015MNRAS.452.1171W

Wilcox H, Nichol RC, Zhao G-B, Bacon D, Koyama K, Romer AK (2016) MNRAS 462:715. https://doi.org/10.1093/mnras/stw1617, http://adsabs.harvard.edu/abs/2016MNRAS.462..715W

Will CM (2014) Living Rev Relativ 17:4. https://doi.org/10.12942/lrr-2014-4, http://adsabs.harvard.edu/abs/2014LRR....17....4W

Winther HA et al (2015) MNRAS 454:4208. https://doi.org/10.1093/mnras/stv2253, http://adsabs.harvard.edu/abs/2015MNRAS.454.4208W

Winther HA, Ferreira PG (2015) Phys Rev D 91:123507. https://doi.org/10.1103/PhysRevD.91.123507, http://adsabs.harvard.edu/abs/2015PhRvD..91l3507W

Zhao G-B (2014) APJS 211:23. https://doi.org/10.1088/0067-0049/211/2/23, http://adsabs.harvard.edu/abs/2014ApJS..211...23Z

Zhao H, Macciò AV, Li B, Hoekstra H, Feix M (2010) ApJL 712:L179. https://doi.org/10.1088/2041-8205/712/2/L179, http://adsabs.harvard.edu/abs/2010ApJ...712L.179Z

Zhao G-B, Li B, Koyama K (2011a) Phys Rev D 83:044007. https://doi.org/10.1103/PhysRevD.83.044007, http://adsabs.harvard.edu/abs/2011PhRvD..83d4007Z

Zhao G-B, Li B, Koyama K (2011b) Phys Rev Lett 107:071303.https://doi.org/10.1103/PhysRevLett.107.071303, http://adsabs.harvard.edu/abs/2011PhRvL.107g1303Z

Part III
Conclusions and Future Work

Chapter 7
Conclusions and Future Work

7.1 A Summary of This Thesis

The concordance model of cosmology, ΛCDM, has undoubtedly withstood the tests of time. In many ways, it is remarkable that this relatively simple model is able to successfully fit and predict a vast range of phenomena in the Universe, such as the temperature fluctuations observed in the CMB, and the large-scale distribution of galaxies. The continuous development of sophisticated numerical and semi-analytic techniques have facilitated tests of this model on non-linear scales where, recently, hydrodynamical simulations within a ΛCDM context have managed to successfully reproduce a large set of observed galaxy properties at low redshift (e.g. Vogelsberger et al. 2014, Schaye et al. 2015).

Despite these successes however, testing the predictions of alternatives to ΛCDM is of vital importance. In the case of CDM, which provides a consistent picture for structure formation on small and large scales, the main source of concern is that despite the many years of targeted direct and indirect detection experiments, the CDM particle has not yet been discovered (see e.g. Arcadi et al. 2017, for a recent review). This, coupled with the non-detection of supersymmetry at the LHC, is gradually narrowing down the parameter space within which traditional CDM candidates are thought to exist. With regards to Λ, studies extending beyond the standard model can be motivated by the fact that the canonical formulation of General Relativity plus a cosmological constant may not be a good description for the nature of gravity on scales beyond the Solar System. Large-scale tests of gravity are particularly timely in anticipation of future surveys like the LSST (Ivezic et al. 2008), DESI (Levi et al. 2013) and EUCLID (Laureijs et al. 2011).

Over the course of this thesis we have examined the nature of structure formation in two possible alternative scenarios: in the first half, we assume that the expansion of the Universe is governed by Λ, but the dark matter is composed of sterile neutrinos rather than CDM. In the second half, the dark matter is assumed to be CDM, but the theory of gravity is modified through the addition of an extra term that depends on

© Springer International Publishing AG, part of Springer Nature 2018
S. Bose, *Beyond ΛCDM* , Springer Theses,
https://doi.org/10.1007/978-3-319-96761-5_7

the Ricci scalar, R, to the Einstein-Hilbert action ($f(R)$ gravity). We summarise the main results of this thesis in the following subsections.

7.1.1 Structural Properties of Sterile Neutrino Dark Matter Haloes

In Chap. 2, we introduced the *Copernicus Complexio* (COCO; Hellwing et al. 2016, Bose et al. 2016) simulations, a pair of dark matter-only simulations in which one volume follows the evolution of structure when the dark matter is CDM, while the other assumes dark matter in the form of a 3.3 keV thermal relic WDM particle. Coincidentally, the linear power spectrum of the thermal 3.3 keV particle is very similar to that of the coldest 7 keV sterile neutrino, a particle whose decay may have been detected in the form of an X-ray line at 3.5 keV (Bulbul et al. 2014; Boyarsky et al. 2014). COCO is amongst the highest resolution N-body simulations of cosmological volumes performed to date, providing unprecedented statistical information about the formation of dark matter haloes and galaxies in these two cosmologies. Both the CDM simulation and its WDM counterpart are run with the same initial phases, allowing comparisons between the two cosmologies on both a statistical and object-by-object basis.

In Chap. 2, we investigated the effects of the characteristic free streaming of WDM particles on the internal structural properties of dark matter haloes. Free streaming of WDM leads to a delay in the average collapse time of haloes below a characteristic mass scale ($\sim 2 \times 10^9 \, h^{-1} \, M_\odot$) compared to CDM, and results in a suppression of the mass function of haloes below the mass scale of dwarf galaxies. We found that while sterile neutrinos reduce the central density of haloes relative to CDM, the density profile preserves the universal NFW form down to the smallest scales resolved in the simulation. We established the evolution of the mass function and the concentration-mass relation as a function of redshift and quantified the spins and shapes of CDM and WDM haloes over seven decades in halo mass. We also provided simple relations describing the dependence of these properties on halo mass and redshift.

In Chap. 3, we shifted our focus from haloes to the substructures of these objects. Interestingly, we found that the radial distribution of WDM subhaloes is almost identical to that in CDM, which is an important result, for example, when comparing the properties of satellite galaxies around the Milky Way. Owing to their lower concentrations at the time of infall, WDM subhaloes with $V_{\rm max} \leq 50 \, {\rm km s}^{-1}$ are more prone to tidal stripping after they are accreted into their host halo.

7.1.2 Galaxy Formation with Sterile Neutrinos

Ultimately, in order to place constraints on the nature of the dark matter, it is necessary to confront the predictions of these models with the data. For this purpose, in Chap. 3 we made use of the Durham semi-analytic model of galaxy formation, GALFORM (Cole et al. 2000; Lacey et al. 2016), to translate the dark matter halo catalogues in COCO into galaxy populations. We found that while many present-day observables show negligible difference between the two models, potentially strong constraints can be made using ultra-faint satellites and the high redshift galaxy population.

A more detailed investigation of both these regimes is performed in Chap. 4, where we apply the Hou et al. (2016) model of GALFORM to range of 7 keV sterile neutrino models with leptogenesis parameters $L_6 = (8, 12, 700)$. While reionisation occurs slightly later in these models than in CDM, the epoch of reionisation in all cases is consistent with the bounds from *Planck*. This can be ascribed to the fact that the bulk of the ionising photon budget is produced by galaxies more massive ($M_\star \sim 10^9 \, M_\odot$) than those affected by the free streaming cutoff in these models. The evolution of the far-UV luminosity functions between $10 > z > 7$ indicates that the high redshift galaxy population builds up more rapidly in the sterile neutrino models than in CDM, which is also reflected in the stellar mass growth rate of bright galaxies. Finally, we also quantified the present-day abundance of Milky Way satellite galaxies and found that the population of ultra-faint dwarf galaxies that may be detected in surveys like DES could potentially rule out the entire family of sterile neutrino particles relevant to the 3.5 keV line.

7.1.3 Elevating Numerical Simulations of f(R) Gravity

In the second half of this thesis, we considered the scenario where the dark matter is CDM, but where the underlying theory of gravity is modified. Specifically, we focussed on the case of the Hu-Sawicki formulation of $f(R)$ gravity (Hu and Sawicki 2007), which is one of the most widely-studied examples of modified gravity theories. In Chap. 5, we validated the widely-employed quasi-static approximation in $f(R)$ gravity, in which it is assumed that the time derivatives of the scalar field are negligible compared to its spatial derivative. We achieved this by rederiving the scalar field equations of motions without making this approximation, and by then performing a series of N-body simulations with increasing resolution with and without the quasi-static approximation. By comparing the non-linear matter power spectra, velocity divergence power spectra and the PDF of the density field, we found that the effects of the scalar field time derivatives are small enough that they can be safely neglected for most practical applications in cosmology. The three models of $f(R)$ gravity we simulated – namely, the F4, F5 and F6 models – span a wide range in the strength of the chameleon screening mechanism, but in all cases the quasi-static approximation is a good one.

In Chap. 6, we introduced a new method for solving the equations of motion in $f(R)$ gravity simulations. The new method relies on a variable redefinition that makes the equations of motion less non-linear, accelerating the rate of convergence of the solution. Having tested our method for a set of high resolution simulations, we found that the new method boosts the performance of the ECOSMOG code (Li et al. 2012) by more than a factor of 20. Importantly, this speed-up is achieved without sacrificing the accuracy of the solution. The method presented in Chap. 6 could, in principle, be applied to other classes of modified gravity theories, and will make it possible to run large volume, high resolution modified gravity setups that would have previously been very expensive to run.

7.2 Looking to the Future

In this thesis, we have studied a limited number of applications of simulations using sterile neutrinos and modified gravity. Before concluding, it is worth pointing out some of the interesting ways in which the investigation in this thesis can be extended, in an effort to place further constraints on these models. Some of the ideas discussed in Sects. 7.2.1 and 7.2.2 have already been published, but they can be used as starting points for more detailed analysis.

7.2.1 *Constraining WDM with Strong Gravitational Lensing*

As we have seen in Chaps. 3 and 4, the largest observable differences between CDM and sterile neutrino models occur at the scale of ultra-faint dwarfs and galaxies at high redshift. However, only limited data are currently available in these regimes. In fact, the starkest difference between CDM and WDM is in the abundance of the dark matter (sub)haloes themselves (Figs. 2.5 and 3.2). Techniques that are able to directly probe the dark matter mass function will therefore provide the cleanest tests for constraining the nature of dark matter. One such method, pioneered by Koopmans (2005) and Vegetti and Koopmans (2009) uses strong gravitational lensing to detect low mass substructures. Briefly, this method uses the fact that the presence of substructures in the central regions of haloes can distort the Einstein ring surrounding a strong lens system. If the (projected) position of the subhalo is in the vicinity of the Einstein ring, it can perturb its surface brightness distribution. Using this method, the authors in Vegetti et al. (2012) reported the detection of a subhalo of mass $1.9 \pm 0.1 \times 10^8\,\mathrm{M_\odot}$ at a significance level of 12σ. Upcoming telescopes such as the SKA and the LSST will substantially increase the sample of strong lens systems, and it is expected that the detection sensitivity will improve to a level that could allow the detection of subhaloes with mass as low as $10^6\,\mathrm{M_\odot}$.

In Li et al. (2016), we performed Monte Carlo simulations of mock strong lensing observations. For the lensing systems themselves, we randomly sampled haloes in the mass range $[10^{13}, 10^{14}]$ h^{-1} $M_{\odot}$ using the mass function from the EAGLE simulations. The abundance and radial distributions of CDM and WDM subhaloes was obtained from the COCO simulations. Using the Monte Carlo simulations, we estimated that approximately 100 strong lens systems with a detection limit of $M_{low} \sim 10^7$ $M_{\odot}$ would be able to clearly distinguish (i.e., $> 2\sigma$) between CDM and a 7 keV sterile neutrino (see Sect. 3.5 for more details). In a follow-up project, Li et al. (2017) found that projected haloes along the line-of-sight dominate the lensing signal, and these intervening objects actually enhance the differences between CDM and WDM. After taking these projection effects into account, the authors find that merely 20 strong lens systems could be enough to distinguish between WDM and CDM at 3σ significance. These results highlight the tremendous potential for strong lensing as a tool for constraining dark matter.

7.2.2 Constraints on WDM Using Observations in the Local Group

Some of the best quality data that are available to us comes from the Local Group, and the situation will improve even further thanks to missions like DES and Gaia (Gaia Collaboration et al. 2016). The star formation histories of dwarf galaxies in the Local Group could be used to probe the nature of dark matter. As we have seen in Fig. 2.4, the collapse time of WDM haloes is delayed below a characteristic mass scale ($\sim 10^9$ h^{-1} $M_{\odot}$ for a 3.3 keV thermal relic). This means that the formation of the first generation of stars in dwarf galaxies is also delayed in WDM compared to CDM, typically by ~ 1 Gyr or so (Calura et al. 2014; Maio and Viel 2015; Governato et al. 2015).

As part of the APOSTLE suite of hydrodynamical simulations (Fattahi et al. 2016; Sawala et al. 2016), in Lovell et al. (2017), we simulated a set of Local Group analogues in 7 keV sterile neutrino dark matter models with lepton asymmetry $L_6 = 10, 120$. For the galaxy formation model, we used the same prescriptions as used by the EAGLE project (Schaye et al. 2015; Crain et al. 2015).

To compare the stellar age distribution of satellites in our Local Group resimulations, we used the following procedure: first, we selected star particles contained within satellites in the mass bin: $\log(M_{\star}/M_{\odot}) = [6, 7]$. Next, we split the stellar ages of the population into three bins sorted by lookback time, t_{lb}: $t_{lb} < 6$ Gyr, 6 Gyr $\leq t_{lb} < 10$ Gyr and $t_{lb} > 10$ Gyr. Figure 7.1 plots the proportion of stellar mass in each of these bins obtained from our simulations, along with the value measured from Local Group dwarf spheroidals compiled by Weisz et al. (2011).

While there is considerable overlap between the three models, some differences can be identified. At least 26% of the stars in all CDM systems are more than 10 Gyr old, whereas four $L_6 = 10$ systems and seven $L_6 = 120$ systems do not meet this threshold. The largest proportion of $t_{lb} > 10$ Gyr stars in $L_6 = 120$ is 41%, younger than seven of the CDM systems. The $L_6 = 120$ symbols are instead clustered towards

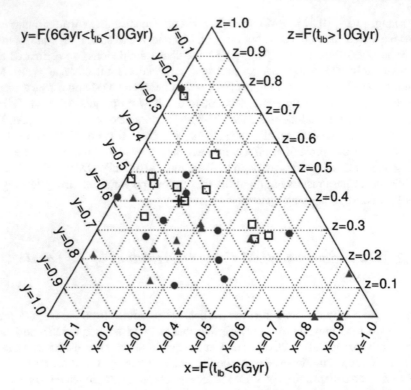

Fig. 7.1 The relative abundances by age of stars in satellites in the stellar mass range $10^6 <$ $M_\star/M_\odot < 10^7$. For each M31 and MW system we add together all of the satellites in the stellar mass bin such that there is one symbol per system: black squares show the results for CDM, while the blue circles and red triangles, respectively, correspond to 7 keV sterile neutrino models with $L_6 = 10, 120$. The x-axis shows the proportion of stars that are younger than 6 Gyr, the z-axis shows the proportion that are older than 10 Gyr, and the y-axis the proportion that are within this age range. The approximate measured values of these quantities for the Local Group dwarf spheroidals as a whole, as presented by Weisz et al. (2011), are shown as the green cross. Figure reproduced from Lovell et al. (2017)

more intermediate age systems, while $L_6 = 10$ systems show a large spread in ages. The value measured for the Local Group is located comfortably within the CDM and $L_6 = 10$ distributions but just outside the $L_6 = 120$ distribution. It is therefore possible that the $L_6 = 120$ 7 keV sterile neutrino produces satellites that are too young compared to the Local Group, though much better resolution is required to confirm this conclusion. Nevertheless, this highlights the potential for age and metallicity distributions of stellar populations in dwarf galaxies as powerful probes of the process of galaxy formation in different models of dark matter.

7.2.3 Confronting Modified Gravity with Data

As we have explained in Chaps. 5 and 6, the inherent non-linearity of modified gravity equations makes these models quite challenging to simulate. For this reason, most studies involving modified gravity have focussed on the differences between these models and the predictions of ΛCDM at the level of dark matter properties only. Barring a few exceptions (e.g. Fontanot et al. 2013, Arnold et al. 2016; He et al. 2016), the conversion of dark matter halo catalogues into an equivalent galaxy population in modified gravity simulations has been limited. This step is necessary in order to make a faithful comparison between the predictions of these models and the data that will be used to constrain them. One way to build such 'mock' catalogues is via the halo occupation distribution (HOD) method (e.g. Berlind and Weinberg 2002, Kravtsov et al. 2004), in which halo catalogues are populated with galaxies by assuming simple functional forms for the average occupation of central and satellite galaxies within haloes. The parameters of the HOD can be calibrated by requiring a match between the number density of galaxies and the projected clustering of galaxies in the mock catalogue and the survey dataset. The mock catalogue can then be made more realistic by taking into account sky completeness, survey masks, redshift selection etc. (see Fig. 7.2).

Once these catalogues have been constructed, the same analysis can be applied to both the mock and the actual survey catalogues to see if any signatures of modified gravity are imprinted on the galaxy distribution. For example, in modified gravity, the presence of a fifth force in unscreened regions boosts the velocities of tracers (subhaloes/galaxies) relative to their counterparts in standard gravity. These differences would be manifest in redshift space and, in particular, in the velocity power spectrum. Since the two-point galaxy clustering does not encode all cosmological information, particularly in modified gravity models (e.g. Hellwing et al. 2013), it is important to study complementary probes such as higher-order moments, topological and morphological characterisations of the galaxy field.

A shortcoming of the HOD treatment is that, by construction, galaxies are assigned to haloes solely based on the host halo mass. As a result, there is no physical information (regarding, say, the local environment the galaxy is due to reside in) encoded in these models. In modified gravity, this environmental information is particularly important as the halo environment may screen the enhanced strength of gravity (or not, as the case may be). Eventually, therefore, the HOD treatment needs to be replaced with a more sophisticated approach like a SAM or hydrodynamical simulations. However, this is easier said than done: constructing a modified gravity SAM is not as straightforward as running an existing SAM on the output of a modified gravity simulation. In many SAMs, a subset of the galaxy formation equations use parameterised versions of the concentration-mass relation, spin distribution etc. of haloes that have been obtained from standard ΛCDM simulations. These relations will be different in modified gravity (e.g. Shi et al. 2015), and will likely take on a more complex form when one takes into account the environmental dependence of the strength of gravity. Building SAMs or subgrid prescriptions tailored to modified

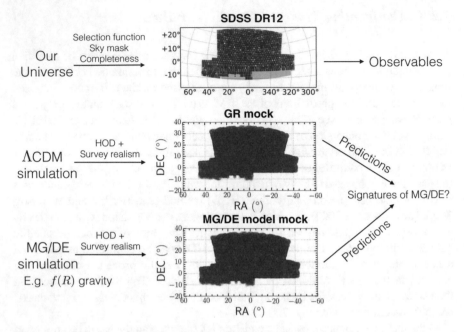

Fig. 7.2 Steps involved in building realistic mock catalogues. The top panel represents the actual survey mask for SDSS DR12 (obtained from Reid et al. 2016); the middle and lower panels are created from ΛCDM and $f(R)$ gravity simulations respectively. In both cases, the catalogues are created using the HOD method, and filtered with the DR11 selection function, sky completeness, redshift selection etc

gravity is therefore a challenging task, but one that could prove to be very rewarding. We hope that the speed-up method presented in Chap. 6 will, for the first time, make it feasible to run large volume hydrodynamical simulations of such models with the high resolution that is necessary.

7.3 Concluding Remarks

It is an incredibly exciting time to be cosmologist. The explosion of data, both on the scales of the faintest galaxies in the local Universe, as well as on the largest, cosmological scales means that we currently have more information in our hands about our Universe than at any time previously. Accurate and detailed theoretical predictions of the standard and non-standard cosmological models are therefore necessary to best interpret what these new datasets have to reveal about the Universe: the nature of the dark matter and dark energy, the assembly of the cosmic web, and the physics of how galaxies form within it. The high precision data that will be delivered by DES, Gaia, LSST, SKA, EUCLID etc. will be exactly what is needed to stress-test the ΛCDM model. In this thesis, we have put forward the case for two popular alternatives to the standard model, in the form of sterile neutrinos as a candidate for the dark matter, and $f(R)$ gravity as an extension of General Relativity. We hope that the content presented in this thesis highlights the prospects for constraining these models further, with a view to one day revealing the true nature of dark matter and dark energy.

References

Arcadi G, Dutra M, Ghosh P, Lindner M, Mambrini Y, Pierre M, Profumo S, Queiroz FS (2017). arXiv:1703.07364

Arnold C, Springel V, Puchwein E (2016) MNRAS 462:1530. https://doi.org/10.1093/mnras/stw1708, http://adsabs.harvard.edu/abs/2016MNRAS.462.1530A

Berlind AA, Weinberg DH (2002) ApJ 575:587. https://doi.org/10.1086/341469, http://adsabs.harvard.edu/abs/2002ApJ...575..587B

Bose S, Hellwing WA, Frenk CS, Jenkins A, Lovell MR, Helly JC, Li B (2016) MNRAS 455:318. https://doi.org/10.1093/mnras/stv2294, http://adsabs.harvard.edu/abs/2016MNRAS.455..318B

Boyarsky A, Ruchayskiy O, Iakubovskyi D, Franse J (2014) Phys Rev Lett. https://doi.org/10.1103/PhysRevLett.113.251301, http://adsabs.harvard.edu/abs/2014PhRvL.113y1301B113,251301

Bulbul E, Markevitch M, Foster A, Smith RK, Loewenstein M, Randall SW (2014) ApJ 789:13. https://doi.org/10.1088/0004-637X/789/1/13, http://adsabs.harvard.edu/abs/2014ApJ...789...13B

Calura F, Menci N, Gallazzi A (2014) MNRAS 440:2066. https://doi.org/10.1093/mnras/stu339, http://adsabs.harvard.edu/abs/2014MNRAS.440.2066C

Cole S, Lacey CG, Baugh CM, Frenk CS (2000) MNRAS 319:168. https://doi.org/10.1046/j.1365-8711.2000.03879.x, http://adsabs.harvard.edu/abs/2000MNRAS.319..168C

Crain RA et al (2015) MNRAS 450:1937. https://doi.org/10.1093/mnras/stv725, http://adsabs.harvard.edu/abs/2015MNRAS.450.1937C

Fattahi A et al (2016) MNRAS 457:844. https://doi.org/10.1093/mnras/stv2970, http://adsabs.harvard.edu/abs/2016MNRAS.457..844F

Fontanot F, Puchwein E, Springel V, Bianchi D (2013) MNRAS 436:2672. https://doi.org/10.1093/mnras/stt1763, http://adsabs.harvard.edu/abs/2013MNRAS.436.2672F

Gaia Collaboration et al (2016) A&A 595:A1. https://doi.org/10.1051/0004-6361/201629272, http://adsabs.harvard.edu/abs/2016A%26A...595A...1G

Governato F et al (2015) MNRAS 448:792. https://doi.org/10.1093/mnras/stu2720, http://adsabs.harvard.edu/abs/2015MNRAS.448..792G

He J-h, Li B, Baugh CM (2016) Phys Rev Lett 117:221101. https://doi.org/10.1103/PhysRevLett.117.221101, http://adsabs.harvard.edu/abs/2016PhRvL.117v1101H

Hellwing WA, Li B, Frenk CS, Cole S (2013) MNRAS 435:2806. https://doi.org/10.1093/mnras/stt1430, http://adsabs.harvard.edu/abs/2013MNRAS.435.2806H

Hellwing WA, Frenk CS, Cautun M, Bose S, Helly J, Jenkins A, Sawala T, Cytowski M (2016) MNRAS 457:3492. https://doi.org/10.1093/mnras/stw214, http://adsabs.harvard.edu/abs/2016MNRAS.457.3492H

Hou J, Frenk CS, Lacey CG, Bose S (2016) MNRAS 463:1224. https://doi.org/10.1093/mnras/stw2033, http://adsabs.harvard.edu/abs/2016MNRAS.463.1224H

Hu W, Sawicki I (2007) Phys Rev D 76:064004. https://doi.org/10.1103/PhysRevD.76.064004, http://adsabs.harvard.edu/abs/2007PhRvD..76f4004H

Ivezic Z et al (2008). arXiv:0805.2366

Koopmans LVE (2005) MNRAS 363:1136. https://doi.org/10.1111/j.1365-2966.2005.09523.x, http://adsabs.harvard.edu/abs/2005MNRAS.363.1136K

Kravtsov AV, Berlind AA, Wechsler RH, Klypin AA, Gottlöber S, Allgood B, Primack JR (2004) ApJ 609:35. https://doi.org/10.1086/420959, http://adsabs.harvard.edu/abs/2004ApJ...609...35K

Lacey CG et al (2016) MNRAS 462:3854. https://doi.org/10.1093/mnras/stw1888, http://adsabs.harvard.edu/abs/2016MNRAS.462.3854L

Laureijs R et al (2011). arXiv:1110.3193

Levi M et al (2013). arXiv:1308.0847

Li B, Zhao G-B, Teyssier R, Koyama K (2012) JCAP 1:051. https://doi.org/10.1088/1475-7516/2012/01/051, http://adsabs.harvard.edu/abs/2012JCAP...01..051L

Li R, Frenk CS, Cole S, Gao L, Bose S, Hellwing WA (2016) MNRAS 460:363. https://doi.org/10.1093/mnras/stw939, http://adsabs.harvard.edu/abs/2016MNRAS.460..363L

Li R, Frenk CS, Cole S, Wang Q, Gao L (2017) MNRAS 468:1426. https://doi.org/10.1093/mnras/stx554, http://adsabs.harvard.edu/abs/2017MNRAS.468.1426L

Lovell MR et al (2017) MNRAS 468:4285. https://doi.org/10.1093/mnras/stx654, http://adsabs.harvard.edu/abs/2017MNRAS.468.4285L

Maio U, Viel M (2015) MNRAS 446:2760. https://doi.org/10.1093/mnras/stu2304, http://adsabs.harvard.edu/abs/2015MNRAS.446.2760M

Reid B et al (2016) MNRAS 455:1553. https://doi.org/10.1093/mnras/stv2382, http://adsabs.harvard.edu/abs/2016MNRAS.455.1553R

Sawala T, et al (2016) MNRAS 457:1931. https://doi.org/10.1093/mnras/stw145, http://adsabs.harvard.edu/abs/2016MNRAS.457.1931S

Schaye J et al (2015) MNRAS 446:521. https://doi.org/10.1093/mnras/stu2058, http://adsabs.harvard.edu/abs/2015MNRAS.446..521S

Shi D, Li B, Han J, Gao L, Hellwing WA (2015) MNRAS 452:3179. https://doi.org/10.1093/mnras/stv1549, http://adsabs.harvard.edu/abs/2015MNRAS.452.3179S

Vegetti S, Koopmans LVE (2009) MNRAS 400:1583. https://doi.org/10.1111/j.1365-2966.2009.15559.x, http://adsabs.harvard.edu/abs/2009MNRAS.400.1583V

Vegetti S, Lagattuta DJ, McKean JP, Auger MW, Fassnacht CD, Koopmans LVE (2012) Nature 481:341. https://doi.org/10.1038/nature10669, http://adsabs.harvard.edu/abs/2012Natur.481..341V

Vogelsberger M et al (2014) MNRAS 444:1518. https://doi.org/10.1093/mnras/stu1536, http://adsabs.harvard.edu/abs/2014MNRAS.444.1518V

Weisz DR et al (2011) ApJ 743:8. https://doi.org/10.1088/0004-637X/743/1/8, http://adsabs.harvard.edu/abs/2011ApJ...743....8W

Appendix
Faster Simulations in Modified Gravity: Comparison with the Truncated Approach and Application to the Symmetron Model

A.1 Performance of the Truncation Method in Chameleon Models

In Barreira et al. (2015), the authors proposed a method to speed up N-body simulations of modified gravity models with Vainshtein screening. The speed up in this method is achieved by truncating the Gauss–Seidel iterations of the scalar field above a certain refinement level, and then computing the solution on those fine levels by interpolating from coarser levels. This *approximate* method agrees very well with the results of the full N-body calculation (see Barreira et al. 2015, for details) due to the fact that in Vainshtein screening models, there is a correlation between higher density regions (or, equivalently, higher refined regions in the simulation box) and screening efficiency. Even when the error induced on the fifth force on the refinements is large, it does not propagate to the total gravitational force because the amplitude of the fifth force is small/screened.

In chameleon models, however, the correlation between high density regions and screening efficiency becomes less marked because of the dependence on the environmental density (in Vainshtein models, the screening efficiency depends on the local density only). For example, in $f(R)$ models, a low mass halo in a dark matter void constitutes an example of a highly-refined region (the centre of the halo can be very concentrated) that may not be screened (either by itself or by the low density environment it lives in). It is therefore interesting to determine whether or not the same truncation method, which works well for Vainshtein models, works equally well in chameleon-type theories.

Figure A.1 shows the relative difference of two truncated $f(R)$ simulations to a full (i.e., not truncated) simulation. The simulation box used for this test is the same as the 'Medium res' setup in the main text, but with $f_{R0} = -10^{-5}$ (the so-called F5 model). The result is shown at three different redshifts and the two labelled truncation schemes are as follows. The case $h_c \leq 0.24$ Mpc/h indicates that the scalar field was only explicitly solved on the coarse level, with this solution being interpolated to all finer levels. In the case of $h_c \leq 0.06$ Mpc/h, the scalar field was

© Springer International Publishing AG, part of Springer Nature 2018
S. Bose, *Beyond ΛCDM* , Springer Theses,
https://doi.org/10.1007/978-3-319-96761-5

Fig. A.1 Relative difference in the matter power spectra at $z = 0$, 0.5 & 1 between a full F5 simulation, and two truncated runs where the Gauss–Seidel iterations of the scalar field have been truncated on finer refinement levels (see the accompanying text). The dashed and solid lines, respectively, correspond to less and more aggressive truncation schemes; h_c is the cell size of the first truncated level in each simulation. The shaded grey band represents the 1% error region around zero. Figure reproduced from Bose et al. (2017)

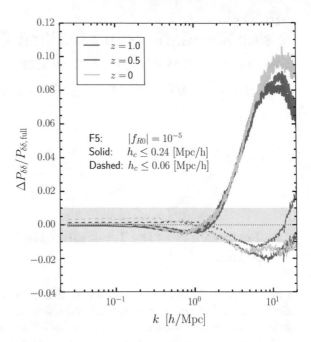

explicitly solved on the coarse, first refinement and second refinement levels, with the solution at the second level being interpolated to all other finer levels. The values 0.24 Mpc/h and 0.06 Mpc/h indicate the cell size of the first truncated level in both these simulations, which ran, respectively, $\approx$10 and $\approx$2 times faster than the full run. For both these truncation criteria, the figure shows that the error can be kept <1% for $k \lesssim 2$ h/Mpc, but for higher modes, it grows to unacceptably large values. For example, at $k \approx 5$ h/Mpc, the error is of $\approx$6%.

The result shown in Fig. A.1 for $f(R)$ should be contrasted with the corresponding picture in the DGP model (which employs Vainshtein screening), in which for the same truncation criteria, the error is always kept below 1% for $k < 5$ h/Mpc (see e.g. Fig. 5 of Barreira et al. 2015). Furthermore, the method described in Chap. 6 results in comparable boosts in performance compared to previous $f(R)$ simulations, but without any loss in accuracy. From this we can conclude that the truncation scheme that works well in Vainshtein screening models is not suitable for chameleon theories.

A.2 Performance of the New Method for the Symmetron Model

As a test of the performance of our new method for other classes of screening mechanisms, we implemented our method for the case of the symmetron model. The code used in this case, Isis (Llinares et al. 2014), is a modified version of RAMSES developed independently of ECOSMOG. Details of the symmetron model and its implementation in Isis are described in Llinares et al. (2014). Briefly, the equation of motion for the scalar field is given by:

$$\nabla^2 \phi \propto (A\rho - 1)\,\phi + \phi^3, \qquad\qquad (\text{Appendix.1})$$

where the quantity A is a function of the parameters of the symmetron model. While the equation is formally equivalent to the $f(R)$ in the main text (Eq. 6.2.2), the screening mechanism operates differently. In the $f(R)$ model, the scalar field screens itself by becoming very massive. On the other hand, in the symmetron model, the screening occurs when a particular symmetry is restored (i.e., when the factor in front of the linear term of the source of Eq. (Appendix.1) becomes positive). Consequently, the model behaves in a different manner to $f(R)$. For instance, negative solutions for the symmetron field, ϕ, are allowed and, thus, the constraints implemented in the $f(R)$ case (Sect. 6.3.2) are not required. We refer the reader to Llinares and Pogosian (2014) for a summary of the complex phenomenology associated with this property of the symmetron field.

The non-linear modified gravity solver in Isis is very similar to that of the $f(R)$ model in ECOSMOG. The code uses an implicit multigrid solver with full approximation storage, which means that the code relies on a Newton–Raphson algorithm to evolve the solution in every step of the Gauss–Seidel iterations. As the discretised equation is cubic, the method proposed in Chap. 6 can be applied in a straightforward manner. As a check of the accuracy of the new method in solving the symmetron field equations, we have repeated satisfactorily the static test presented in the original Isis paper (Fig. 2 in Ref. Llinares et al. 2014). However, we find that there is no major difference in the performance of the standard Isis implementation compared to using the new method, either in terms of the run time, or the convergence rate of the iterative solver.

In order to gauge the difference in computing time between the old and new methods for the symmetron model, we have run a set of five different realisations of a box of size 150 Mpc/h on a side, containing 256^3 particles. For each realisation, there are three sets of simulations: ΛCDM and the symmetron model using the old and new methods. Overall, we do not find any improvement in the performance of Isis using the new method. For both the old and new methods, the overhead compared the ΛCDM simulation is of the order of $\sim$170% and, in fact, the run time using the new method is actually $\sim$1% slower than using the default implementation - this is explained by the fact that $\sim$1% more iterations were required for the whole set of five realisations using the new method. The convergence criterion on the residual was set to 10^{-6} for both symmetron runs; we find that, unlike in the $f(R)$ model, making the convergence criterion even stricter does not impact the run time of the symmetron simulations by a great amount.

The reason why the performance of the code appears to be insensitive to the details of the iteration scheme is seemingly related to the type of screening mechanism used by the symmetron model. The symmetron mechanism is based on a density threshold above which the solution very quickly approaches zero and thus decouples the scalar field from matter. This makes the solutions more stable and, therefore, not strongly dependent on the details of the solver employed. Since the default solver in Isis does not involve a non-linear change of variables to force a stable, positive solution (as in the $f(R)$ case), the performance is already similar to what ECOSMOG can do for $f(R)$ using the new method.

References

Barreira A, Bose S, Li B (2015) JCAP 12:059. https://doi.org/10.1088/1475-7516/2015/12/059, http://adsabs.harvard.edu/abs/2015JCAP...12..059B

Bose S, Li B, Barreira A, He J-h, Hellwing WA, Koyama K, Llinares C, Zhao G-B (2017) JCAP 2:050. https://doi.org/10.1088/1475-7516/2017/02/050, http://adsabs.harvard.edu/abs/2017JCAP...02..050B

Llinares C, Pogosian L (2014) Phys Rev D 90:124041. https://doi.org/10.1103/PhysRevD.90.124041, http://adsabs.harvard.edu/abs/2014PhRvD..90l4041L

Llinares C, Mota DF, Winther HA (2014) A&A 562:A78. https://doi.org/10.1051/0004-6361/201322412, http://adsabs.harvard.edu/abs/2014A%26A...562A..78L

Index

© Springer International Publishing AG, part of Springer Nature 2018
S. Bose, *Beyond ΛCDM* , Springer Theses,
https://doi.org/10.1007/978-3-319-96761-5